HUMAN FACTORS CONSIDERATIONS
OF
UNDERGROUNDS IN INSURGENCIES

PRIMARY RESEARCH RESPONSIBILITY:

Andrew R. Molnar

With research collaboration of:

Jerry M. Tinker

and

John D. LeNoir

Research and writing completed: 1 December 1965

SPECIAL OPERATIONS RESEARCH OFFICE
The American University
Washington, D. C.

Operating Under Contract With
The Department of the Army

Reproduction in whole or in part is permitted for any purpose of the United States Government.

This research is accomplished at the American University by the Special Operations Research Office, a nongovernmental agency operating under contract with the Department of the Army.

The content of this publication, including any conclusions or recommendations, represents the views of SORO and should not be considered as having Department of the Army approval, either expressed or implied.

Published by Conflict Research Group

First published by Special Operations Research Office in 1966

ISBN: 978-1-925907-17-9

CONFLICT
RESEARCH
GROUP

FOREWORD

The dynamics of social change are of prime concern for many social scientists. However, in the desire to understand the broad characteristics and societal impact of revolutionary movements, we often neglect the study of the human element involved in them.

Those ideologists who write revolutionary dogma or those who report the history of great revolutions speak of the masses as if they were a living being. But what of the individuals that make up the mass? What are the wants and aspirations that lead individuals to join, to remain in, or to leave these underground movements? How are law-abiding citizens tempted to the dangerous life of the underground? And once committed, what influences them to stay? What rules of behavior and decision enable them to survive such a hazardous existence?

To understand the individual, his reasons, his behavior, and the pressures that society places upon him is at the heart of the problem of social change. The battleground of insurgency has been described as the hearts and minds of men. There the understanding of the human element is basic to understanding the dynamics of social change.

The information synthesized in this report is but an initial step in the attempt to understand the motivation and behavior of those in underground and insurgent movements.

Comments of readers are most welcome.

PREFACE

Human Factors Considerations of Undergrounds in Insurgencies is the second product of SORO research on undergrounds. The first, *Undergrounds in Insurgent, Revolutionary and Resistance Warfare*, was a generalized description of the organization and operations of underground movements, with seven illustrative cases. The present study provides more detailed information, with special attention to human motivation and behavior, the relation between the organizational structure of the underground and the total insurgent movement, and Communist-dominated insurgencies.

Because an understanding of the general nature of undergrounds is necessary to more detailed considerations, some of the information from the earlier study of undergrounds has been included in this report. Wherever possible, material from insurgency situations since World War II has been used. Occasionally, however, it was necessary to use information from studies of World War II underground movements in order to fill gaps about certain operations.

In the methodological approach it was assumed that confidence could be placed in the conclusions if data on underground operations and missions and similar data could be found in other insurgencies. An attempt was made to base conclusions on empirical information and actual accounts rather than theoretical discussions, and upon data from two or more insurgencies. An effort was made to find internal consistencies within the information sources. For example, if units were organized and trained to use coercive techniques for recruiting, and defectors described having been recruited in this manner, the conclusion that people were coerced into the movement can be made. Because of this approach there is a good deal of redundancy within and among the various chapters.*

While the main emphasis in this report has been on underground organization, many characteristics can be understood only in relation to overt portions of the subversive organization. Therefore, discussions of guerrilla forces, the visible outgrowth of undergrounds, and of Communist structures, which often inspire, instigate, and support subversive undergrounds, have been included. The report is designed to provide the military user with a text to complement existing training materials and manuals in counterinsurgency and unconventional warfare, and to provide helpful background information for the formulation of counterinsurgency policy

---

* See appendix A for the methodology used in this study.

and doctrine. As such, it should be particularly useful for training courses related to the counterinsurgency mission.

The authors wish to express thanks to a number of persons whose expertise and advice assisted substantially in the preparation of this report. Mr. Slavko N. Bjelajac, Director of Special Operations for the Office of the Deputy Chief of Staff for Operations, Department of the Army, on the basis of his personal experiences and special interest in the study of underground movements, contributed guidelines and concepts to the study.

Four men reviewed the entire report: Dr. George K. Tanham, Special Assistant to the President of the Rand Corporation of Santa Monica, California, made many helpful suggestions based upon his firsthand experiences and study of Communist insurgency; Dr. Jan Karski, Professor of Government at Georgetown University, Washington, D.C., whose personal experience as a former underground worker is combined with a talent for thorough, constructive criticism, also helped the final manuscript; Dr. Ralph Sanders of the staff of the Industrial College of the Armed Forces, Washington, D.C., offered a careful and useful critique of the manuscript and helpful suggestions; Lt. Col. Arthur J. Halligan of the U.S. Army Intelligence School, Fort Holabird, Maryland, provided valuable suggestions based upon his experience in Vietnam.

Within SORO, Dr. Alexander Askenasy, Brig. Gen. Frederick Munson (Ret.), Mr. Phillip Thienel, Mr. Adrian Jones, Dr. Michael Conley, Mrs. Virginia Hunter, and Mrs. Edith Spain contributed to the end product.

CONTENTS

LIST OF ILLUSTRATIONS

LIST OF MAPS AND CHARTS

Figure

LIST OF TABLES

SUMMARY

PURPOSE

The objective of this study is to describe, on the basis of existing empirical information and current state of knowledge, the organizational, motivational, and behavioral characteristics of undergrounds in insurgent movements and to relate these characteristics to the total revolutionary structure, mission, and operations.

Particular emphasis has been placed on—

(1) Describing underground organizations and relating them to the total insurgent organization.

(2) Describing Communist uses of undergrounds and their role in Communist-dominated insurgencies.

(3) Summarizing descriptive and empirical information on motives for joining, staying in, and defecting from underground organizations.

(4) Describing certain underground administrative, psychological, and paramilitary operations, and noting the human factors which appear to be related to their success or failure.

(5) Describing the organization and countermeasures used by governments to suppress or eliminate undergrounds.

SYNOPSIS

PART I. ORGANIZATION

The structure of an insurgent or revolutionary movement is much like an iceberg. It has a relatively small visible element (the guerrilla force) which is organized to perform overt armed operations, and a much larger clandestine, covert force (the underground). The underground carries on the vital activities of infiltration and political subversion; it establishes and operates shadow governments; and it acts as a support organization for the guerrillas.

An insurgent or revolutionary movement is defined as a subversive, illegal attempt to weaken, modify, or replace an existing governing authority through the protracted use or threatened use of force by an organized group of indigenous people outside the established governing structure.

An "underground" is defined as the clandestine, covert element of the insurgent movement. In the initial stages of an insurgency, the entire organization functions in a covert manner and is therefore underground. A long, careful preparation is required, with the underground exploiting dissatisfaction and discontent, to create a structure strong enough to support a specialized organization for armed activity. Eventually, a dual structure is formed. One element conducts overt guerrilla activities while the other, the underground, continues its infiltration and political subversion, intensifying its shadow-government activities.

The guerrilla organization, when fully developed, is composed of a mobile main force structured along conventional military lines, and two paramilitary forces—a regional force and a local militia—which conduct limited actions and support the main force. The underground arm of the insurgent movement is usually pyramidally structured, from a broad base of cells, through branches, districts, states or provinces, to a national headquarters at the top.

An essential feature of most underground organizations is compartmentalization, designed to protect the organization's security. The cellular structure follows the underground "fail-safe" principle: if one element fails, the consequences to the whole organization will be minimal. The degree of compartmentalization and the number of cells established depend upon the size of the organization, the degree of popular support given to government forces, and the relative danger that security forces pose.

Generally, there are three types of underground cells: 1) the operational cell, usually composed of a leader and a small group of members who operate directly as a unit; 2) the intelligence cell, which undertakes espionage, infiltration, and intelligence-gathering activities and is highly compartmentalized, its leader directing his agents through an intermediary; and 3) the auxiliary cell, composed of part-time workers, often found in front groups. Parallel cells are often established to provide backup support in case primary cells are compromised. Another underground structure consists of a series of interlocking cells to carry out functions which require a division of labor, such as manufacturing weapons, acquiring supplies, and providing for escape and evasion. Cell size usually varies from 3 to 8 members.

To show how the organizational structure of undergrounds changes in protracted revolutions, it is useful to categorize phases in the evolution of conflict. The first phase is the *clandestine organization phase* in which the underground begins developing such administrative operations as recruiting, training cadres, infiltrating key government organizations and civil groups, establishing escape-and-evasion nets, soliciting funds, establishing safe areas, and developing external support. During this phase, cell size is kept small and the organization is highly compartmentalized.

The second phase is marked by a subversive and *psychological offensive* in which the underground employs a variety of techniques of subversion and psychological operations designed to add as many members as possible. Covert underground agents in mass organizations call for demonstrations and, with the aid of agitators, turn peaceful demonstrations into riots. Operational terror cells carry out selective threats and assassinations.

In the third or *expansion phase* the organization is further expanded and mass support and involvement are crystallized. Front organizations and auxiliary cells are created to accommodate and screen new members.

During the *militarization phase* overt guerrilla forces are created. Guerrilla strategy usually follows a three-stage evolution. In the first stage, when guerrillas are considerably outnumbered by security forces, small guerrilla units concentrate on harassment tactics aimed at forcing the government to overextend its defense activity. The second stage begins when government forces are compelled to defend installations and territory with substantially larger forces. The third stage marks the beginning of the full guerrilla offensive of creating and extending "liberated areas." During all of these stages the underground acts as the supply arm of the guerrillas, in addition to carrying out propaganda, terrorist, sabotage, and other subversive activities. Crude factories are set up by the underground and raids are conducted to obtain supplies and weapons. Caches are maintained throughout the country and a transportation system is established. Finances are collected on a national and international basis. Clandestine radio broadcasts, newspapers, and pamphlets carry on the psychological offensive. The underground continues to improve its intelligence and escape-and-evasion nets.

In the fifth phase, the *consolidation phase*, the underground creates shadow governments. Schools, courts, and other institutions are established to influence men's minds and control their actions, and covert surveillance systems are improved to insure positive control over the populace.

Although there are general similarities in the organization and development of all underground movements, the Communists have added important refinements to the strategy of revolutionary warfare. With the establishment of the Comintern, an international dimension was added to organization for subversion.

Since the days of Lenin, the Communists have stressed that successful revolutions must be led by professional Communist elites. Consequently, the party structure remains the controlling force throughout the insurgency, expanding and reorganizing as necessary.

Typically, the Communist Party has both open and covert organizations, based on cells and performing in a conspiratorial manner. The lower bodies of the party select representatives to serve on higher committees until, at the top, a central committee is formed. In turn, all lower committees are responsible for carrying out the decisions of the higher ones. In theory, each proposal is discussed at the lower levels of the party and representatives pass the decisions to higher levels until a final decision is made at the top. In practice, decisions generally flow from top to the bottom, with the lower levels permitted *pro forma* discussion of the decisions. Also in theory, the party is organized on the principle of collective leadership with all decisions agreed upon by the majority of officers at any particular committee level. In practice, however, it functions in a highly centralized manner, with all authority and command decisions coming from the top.

Institutionalized criticism and self-criticism sessions are characteristic of Communist organizations. The criticism sessions increase the efficiency of the party by subjecting its operations to constant review and revision, and by leading to normative behavior among its members.

Operating through mass organizations and "united fronts," Communists have found that a small group of highly disciplined individuals can achieve maximum effectiveness. The objective in infiltrating organizations with a mass character is to neutralize agencies which support the government, justify the insurgent cause, and mobilize mass support. In a united front, the

Communists seek to consolidate and unite forces of discontent against the government, as well as to gain access to and control of groups not identified with the Communist position. The Communists have utilized the united front technique in most insurgencies, forming alliances with other political groups by offering them the organizational support of the Communist Party.

Front groups are used when the Communists are unable to infiltrate existing organizations or unite them in a mass front. These organizations usually espouse some worthy cause in order to get the support of respectable citizens, but the leadership remains firmly in the hands of the Communists.

The transition from a peacetime clandestine underground organization into a "national liberation" movement carrying on protracted revolutionary warfare involves a number of major organizational modifications. A dual structure of underground and guerrilla arm is set up, in which the Communists maintain interlocking positions of leadership.

One of the significant functions of the Communist underground is the establishment of "shadow governments." Usually initiated in towns and villages, shadow governments attempt to subvert government control at all levels, particularly at the grassroots level. New political institutions, instruments of control, and symbols of authority are created. Population control is maintained through multiple organizational membership and techniques of agitation and propaganda. Instruments of social force, such as courts and law enforcement agencies, are also used to coerce the doubters.

PART II. MOTIVATION AND BEHAVIOR

Various environmental factors are cited as having a relationship to the "causes" of insurgency. It has been suggested that a nation's stage of economic development, rural-urban composition, rate of illiteracy, or educational level, affect the occurrence, if not the outcome, of an insurgency. However, in a review of 24 insurgencies since 1946, it was found that none of these factors was related to either the outbreak or the outcome of insurgency. For instance, a country's relative stage of economic advancement, as measured by its gross national product per capita, had little relationship to the occurrence or the success or failure of insurgency. While economic factors may be important in the context of local or regional situations, the gross national product is not a predictive indicator of incipient insurgency.

Insurgency involves only a small minority of a country's population as active participants and can be described as a low-intensity conflict. Most of the participants are members of the underground and perform their normal functions within the society along with their clandestine, covert activities. A number of characteristics of insurgency members have been identified. Not surprisingly, men have been found to constitute the majority of both the underground and guerrilla organizations, although women have been active in both organizations, especially in the underground. Both in age and occupation, members reflect the general makeup of the country.

The *motivation* for joining an underground movement is typically complex. Usually, persons join because of a combination of interrelated factors, most frequently personal and situational in nature. Ideological or political reasons seem to have inspired only a small percentage, and propaganda promises appear to have had little effect. Although coercion alone is only a small factor, coercion coupled with other positive incentives is a significant factor. Government persecution, real or imagined, also leads people to join the insurgents.

An insurgent's motives for remaining in the underground seem often to be quite different from his motives for joining. He develops loyalties toward friends and comrades, or may be influenced by indoctrination and other propaganda. Close surveillance and threats of retaliation often make it difficult to withdraw from the movement or to defect to the government forces. Simple inertia may keep him in the movement.

An insurgent may withdraw either by ceasing to participate in the movement or by defecting to the government side. Once he is disaffected he seeks the easiest and safest avenue of escape. If circumstances are such that he can simply leave, he is likely to do that. If the possibility of defecting arises first and it is relatively easy and safe, he may defect; only defection to the government cause is recordable. Insofar as the guerrilla part of the movement is concerned, situational and personal factors are more often involved than ideological ones in a decision to withdraw or defect. Once the individual is disaffected, he usually begins to rationalize and finds many flaws in the goals, organization, or individuals involved in the movement. Government appeals and offers of rehabilitation programs, when known, tend to be an influencing factor in defection.

Ideology is an important factor in unifying the many divergent interests and goals among an underground movement's membership. As a common set of interrelated beliefs, values, and norms, ideology is used to manipulate and influence the behavior of individuals within the group. Ideology also offers a way for individuals to reduce the ambiguity and uncertainty in their social and physical environment and give meaning and organization to unexplained events.

Group membership serves to satisfy several types of individual needs: patriotism, the sense of "belonging," recognition, and enhancement of self-esteem. Strong organizational ties protect an individual from external threats and offer him an opportunity to achieve economic or political goals not otherwise attainable. Group membership does a great deal to condition and mold an individual's behavior. For example, group membership in an underground provides a set of standards, so that an individual always knows implicitly what is right or wrong, what can or cannot be done. Underground membership structures and narrows an individual's exposure to perception of his environment. Because his view of life, of events, and of news is colored by his feelings and behavior, group organization also conditions attitudes and perceptions.

A variety of factors affect the degree of influence underground membership exercises over individuals. Small cells or working groups exercise more effective control than large ones. Frequency of meetings and length of membership affect the development of intimate relationships. The more highly structured the underground and the more clearly defined the relationships and duties, the greater the influence exerted.

Underground movements have been described as "normative-coercive" organizations, employing both persuasive group pressures and overt coercion. They are normative in that institutional norms and mores secure behavioral conformity to certain rules and group membership satisfies certain individual needs and desires. However, coercive power is applied through the threat or application of physical sanctions, or through the deprivation of certain satisfactions.

The Communists, through emphasis on ideology, democratic centralism, criticism and self-criticism techniques, the committee system and the cell structure, have created a high degree of cohesiveness. Furthermore, their techniques seem to be effective in providing informational feedback to the leadership. The criticism–self-criticism sessions particularly serve to reinforce those normative patterns of behavior established during indoctrination.

Clandestine and covert behavior is an important feature of underground practice. By establishing behavior patterns that avoid drawing attention to the underground movement, the underground organization is protected from detection. By appearing normal and inconspicuous, the underground member makes it difficult for security forces and other citizens to detect his membership in a subversive organization.

Clandestine behavior consists of actions in which the underground member endeavors to conceal his involvement. Covert behavior attempts to conceal and cover underground activities from observation. Various techniques have been employed by the underground to achieve secrecy. Organizational practices, such as cellular structures, false fronts, and false records and communications, disguise underground operations. Certain covert communication practices such as disguised couriers, mail-drops, and various signals are also used. Members of the underground also capitalize upon customs and norms which people accept without question, or play upon human susceptibility to authority or suggestion, to effectively disguise its operations and evade police interference.

PART III. ADMINISTRATIVE OPERATIONS

Effective underground administrative operations are essential for the survival and expansion of an insurgent movement; unless its recruitment, training, and financial needs are serviced, it can neither function nor grow. Operating in a potential or actual hostile environment, underground organizations face the requirement of balancing the need for cautious, normal administrative functions against the risks and vulnerabilities that inevitably accompany any aggressive action. To achieve this, underground leaders must adapt administrative techniques to the changing, but always risky, situations of insurgency.

In underground *recruitment*, for example, the means as well as the kind of individual recruited depend upon the movement's stage of development and the political-military situation. During the early phase, primary attention is given to selecting a well-disciplined cadre. The essential need is for tight security; hence, recruitment is highly selective and recruits are thoroughly screened. Various tests and oaths are required of recruits to commit them to the movement and confirm their reliability. In later phases, as the insurgency gains in organizational sophistication, emphasis is placed on expanding the size of the multiple elements of the movement and increasing its mass support from outside. Through either persuasion or coercion, the original underground organization attempts to create a parallel mass organization. Persuasion comes through propaganda or programs to assist the people, such as helping villagers harvest the crops, build schools, etc. Once a feeling of indebtedness is created, the underground asks for help in return and may recruit or "draft" men. Coercion may vary from simple "armed invitation" and the impressment of "volunteers," to methods of alienating or compromising an individual vis-à-vis the government so that he has no other alternative but to join or support the underground. Indirect techniques of mass recruitment include group pressure and suggestion. The use of indigenous "keymen" within a town or an organization is another effective underground technique. Appeals to join the underground capitalize upon the love of power, pressure from friends, anticipation of future rewards, hatred, or ideology and patriotism.

The *training* of special underground cadres is an essential corollary to recruitment. The underground seeks to maximize its effectiveness by preparing recruits in techniques of clandestine behavior, agitation, subversive activities, terror, sabotage, intelligence methods, and guerrilla warfare. Indeed, many underground movements have established special schools to give recruits both practical and ideological training. The international Communist movement has long stressed the essential role of training, and its schools—from the Moscow Lenin School of the 1920's to the Castro-Cuban training camps of the 1960's—provide some of the best examples of education tailored to support subversion. They have sought not only to prepare trainees in the art of underground and guerrilla tactics, but to imbue them with a sense of dedication and ideological purpose to insure their carrying out directives even when the leadership has no direct control. To support this kind of training, literature such as Lenin's "What Is To Be Done?" or Mao Tse-tung's writings are used because they provide an essential link between the practical and the ideological.

Underground *finance* is another essential element of insurgent administrative operations. The underground may tap external sources, such as foreign governments or fraternal groups, or they may raise funds within their own country. They may persuade people to give voluntary contributions, or they may make legal or illegal sales of goods. If voluntary sources are inadequate, the underground frequently resorts to coercive methods, such as robberies, extortion, or, in areas they control, imposition of taxes.

People contribute to the underground for a variety of reasons: ideological allegiance to the cause, social pressure, present or future protection, chance of personal gain, or a desire to be on the winning side.

PART IV. PSYCHOLOGICAL OPERATIONS

In an insurgency, neither victory nor warfare can be conceived solely in military terms. Few insurgencies have been won or lost by large, decisive military battles. Usually some combination of military, political, and social means is used. Much of the political leverage involved in favorable settlements is derived from effective underground psychological operations. Through the techniques of psychological operations the underground attempts to produce a social-political climate favorable to its control. To the underground, and especially the Communist underground, influencing opinions and attitudes is not an end in itself, but a means to enhance their organizational work among broad elements of society. Favorable attitudes and good intentions alone do not create revolutions: organization is necessary for action.

Propaganda and agitation (in Communist jargon called "agitprop") are the principal forms of underground psychological operations. Propagandists and agitators identify their appeals with society's recognized values so as to entice those who accept these widely held views to accept the underground. The tools of the underground propagandist include most techniques of the mass media—newspapers, leaflets, radio—and also stress word-of-mouth communication. It is here that the agitator in Communist movements plays a central role: it is his task to overcome the inevitable barriers in communication and to see that the message reaches the target audience in a credible and meaningful form. Appeals are usually emotional and may take the form of threats. They are directed at self-interest and prejudice. The agitator must not only convince his audience but must convert attitudes into mass action, dislodge complacency, and intensify dissatisfaction. He encourages his audience to respond and provide feedback, and uses group beliefs, values, and norms to win support through social pressure.

To further alienate or crystallize public opinion against a government, the underground may advocate and organize *passive resistance*. Passive resistance implies a large, unarmed group whose activities capitalize upon existing social norms in order to provoke action by government security forces that will serve to alienate large segments of public opinion. The passive resister seeks to persuade the populace to withdraw its support and cooperation from a government; his weapon in this persuasion is his ability to suffer—to martyr himself—and demonstrate by his suffering that the government is tyrannical and unfit.

Passive resistance takes various forms and employs various nonviolent techniques. It may use attention-getting devices such as demonstrations, mass meetings, picketing, or, at another level, techniques of noncooperation such as absenteeism, and civil disobedience. It is difficult for security forces to effectively control passive resisters. The effectiveness of passive resistance, particularly civil disobedience, rests on securing widespread compliance. In gaining this popular support, passive resistance provides strong social coercion to influence the undecided or uncommitted to join the underground movement.

The underground, however, seldom relies solely on the attractiveness of its appeals or on the persuasiveness of its goals. When other techniques of psychological operations fail, it brings coercive means to bear. *Terrorism* represents a strong negative sanction to ensure that recalcitrant individuals comply with the underground's demands. Terrorism is used to support other underground efforts such as propaganda and agitation, and is always used with an understanding of its psychological effects and potential.

Terror may best be described as a state of mind that varies in effect and degree among individuals. It captures the attention of the individual and makes him aware of and vulnerable to the terrorists' demands. The utility of terrorism for an underground movement is multifarious: it may be used to disrupt government control of the population; it may demonstrate underground strength and attract popular support; it may suppress cooperation with the government by "collaborators"; and may be used to protect the security of the clandestine organization.

Three types of terrorism can be distinguished: unorganized, support, and specialized. In spite of rules against unsanctioned acts of terror, they do occur. These acts are unorganized and are committed by groups or individuals during underground operations. Support groups, however, are sanctioned to enforce underground directives and threats through the use of terrorism. For selective targets, specialized terror units made up of "professionals" are employed.

Another technique used by the underground to alienate the populace from governments is the *subversive manipulation of crowds*. The crowds that participate in civil disturbances are particularly vulnerable to manipulation by a relatively few underground agitators who direct them toward emotional issues and arouse them against authority. Usually a subversively manipulated civil disturbance evolves in four phases: 1) the *pre-crowd phase*, when the subversive elements organize, train, and plan for their action; 2) in the *crowd phase*, a group largely composed of individuals who have been conditioned either by subversive manipulation or other events is assembled; 3) the *civil disturbance phase*, when agents maintain emotional excitement, create martyrs, and focus the riot situation; and 4) the *post-disturbance phase*, when the emotions aroused are capitalized upon in calling of strikes, spreading of violence, or creating united front parties and pressure groups.

In the subversive manipulation of crowds, as when dealing in propaganda, agitation, passive resistance, and terrorism, psychological operations are concerned not only with the "objective" world about an individual but with the world as seen by the individual. Although the "real" world or the "facts" are important in psychological operations, what matters most is what people believe and can be made to believe. The intent of the underground in crowds and riots is to focus, direct, manipulate, and create beliefs which will crystallize support for the underground.

PART V. PARAMILITARY OPERATIONS

The underground performs a variety of paramilitary activities. The political and armed activities of an insurgency overlap both in function and in personnel. Usually inferior in numbers and resources to the government security forces, the underground must use every opportunity and capitalize upon every advantage in undertaking paramilitary operations. This requires careful planning of underground missions involving the development of contingency plans, rehearsals of the mission in advance, and careful study of enemy vulnerabilities. Techniques used by undergrounders to exploit vulnerabilities in planning missions include infiltration, surprise, deception, diversion, and creation of fatigue through continuous harassment and provocation. Many other factors are also considered by the underground leaders in planning missions, from situational factors such as the most strategic time of day to human factors of enemy morale and confidence.

Adequate *intelligence* estimates are, of course, prerequisite to effective planning of underground missions. Intelligence allows underground operations planners to establish the necessary priorities among enemy targets and to expose, create, and take advantage of security vulnerabilities. Intelligence is also critical in the planning of psychological operations; it reveals the attitudes, grievances, and specific problems of a target group so that propaganda themes and agitation slogans may be appropriately tailored. Indeed, one of the first tasks facing an underground movement is the establishment of an adequate intelligence network, most frequently on a cellular basis. Underground intelligence relies on both reconnaissance and on the cooperation of the "part-time" members in towns or villages. The use of "innocents"—children, old men, and women—is particularly common in providing intelligence about the movement of government security forces.

The most common paramilitary operations are *ambushes and raids*. Because ambushes involve a surprise attack from a concealed position on terrain of the attackers' choosing, they are as popular and easy to launch as they are devastating and difficult to counter. Not surprisingly, the ambush plays a part in 60 to 70 percent of Communist armed action. Raids and ambushes are useful to the underground in acquiring weapons and supplies, harassing and demoralizing government forces, delaying or blocking the movement of troops and supplies, destroying or capturing government personnel or installations, and undermining confidence among the populace in the power of government. Tactics in ambushes and raids stress detailed intelligence reports, careful planning, and a boldness of imagination that uses the element of surprise to its best advantage.

Unless security forces obtain complete intelligence on every move of the underground, it is largely impossible to prevent ambushes and raids. However, it is possible to forestall or at least lessen the effectiveness of ambushes. Counterambush strategies usually emphasize that: 1) security forces should not follow consistent patterns of movement; 2) reaction of troops must be swift and automatic and soldiers should be trained to "rush through" an ambush; 3) effective outside communications should be maintained at all times so that reinforcements may be called; 4) ambushers should be aggressively pursued; and 5) rural activity should be carefully observed and intelligence strengthened.

Sabotage is another principal underground paramilitary activity having as its objective the destruction or damage of resources important to the enemy's military effort. In general, underground sabotage falls into two categories: *strategic* and *general*. Strategic sabotage involves hitting targets of key importance by specially trained units in carefully planned missions. General sabotage, on the other hand, is directed at nonstrategic targets with the purpose of encouraging similar acts by the populace, as well as hampering the government in attaining its military capacity. Such acts serve to propagandize the underground movement's strength and popular support as well as to further commit the citizenry to its cause.

Essential to the planning of underground paramilitary operations are methods of *escape and evasion*. Underground escape-and-evasion networks usually consist of established escape routes and hideouts—"safe houses" for temporary stopover or permanent refuge. Care is taken to provide necessary supplies and cover stories for all hideouts. To protect the secrecy of the escape-and-evasion network, all strangers seeking assistance are carefully screened and interrogated. A system of hideouts is also a critical feature of any underground movement's effort to infiltrate outside persons into a country, such as North Vietnamese agents into South Vietnam. Usually such infiltration networks rely upon hideouts in remote, rural areas.

PART VI. GOVERNMENT COUNTERMEASURES

The most effective countermeasure is the use of immediate, overpowering force to repress the first signs of insurgency or resistance. Nations with a representative or constitutional form of government are often restrained from such action by moral, legal, and social considerations, and often attempt to combat the first recognized signs of underground movement through social, economic, or political reforms. All too frequently, however, these positive programs fail, either because of the advanced stage of the underground movement, or because of inadequate resources or time. A government must then organize for more direct, increasingly forceful countermeasures.

As an insurgency gains momentum and government countermeasures move from simple police action to involvement of the armed services, a new centralized command structure is generally required for effective counterinsurgency action. Care must be taken, however, to leave area commanders a certain amount of tactical autonomy to permit swift and aggressive counteraction. Frequently a unified intelligence organization is also established so that intelligence information may be processed rapidly and efficiently, with little duplication of effort. The multiple system of intelligence organization, in which a number of separate intelligence groups work simultaneously, has the advantage of being less vulnerable to compromise by underground infiltration than the unified type.

The character of modern underground and guerrilla activity has added a new dimension to *intelligence* functioning in counteraction. In counter-insurgent warfare the enemy is elusive and targets are transitory. As a consequence, rapid response to intelligence is of crucial importance. Also, the kind of intelligence materials required for action are different. Counter-insurgency intelligence must provide long-range intelligence on the stable factors in the insurgent situation, such as demographic factors, nature of the underground organization, characteristics of those recruited, and the kinds of appeals made, as well as short-run information on specific individuals—biographies of underground suspects, their families, contacts—and on the behavior patterns of the underground. In counterinsurgency much intelligence, particularly contact intelligence in the rural areas where the underground thrives, is based upon informants—either paid, voluntary, or infiltrated agents. Cordon-and-search operations have frequently been used in gathering intelligence where the populace does not cooperate for fear of reprisal from the underground. Surveillance and interrogation provide another source of intelligence.

Defection programs have also played a significant role in the outcome of several counterinsurgency efforts. The psychological impact of defection on other members of the underground is significant and, in addition, defectors may provide considerable intelligence data. Defectors usually decide to defect because of situational factors—from certain long-range factors such as an estimate of the probable outcome of the insurgency, as well as from such short-range factors as disagreement with superiors or adverse living conditions. Because many undergrounders and guerrillas are coerced to join the movement, or join because of highly specific grievances, they can be persuaded to defect if they can be convinced that they will receive good treatment. At the point of decision, the defector is most concerned about the future and fear of possible retaliation. The government's goal should be to communicate with potential defectors, telling them of safe systems and known procedures for defecting. In organizing defection programs, a concerted effort must be made to coordinate psychological operations with other programs. If the government says that defectors will be given fair treatment and then government soldiers or police shoot or punish men who surrender, confidence in the government's promises will obviously be diminished.

Population control is an essential feature of counterinsurgency action. It seeks to accomplish two different, yet integrally related, countermeasure objectives: to restrict the movement of the underground and to separate it both physically and psychologically from the populace. The principal techniques of population control are collective-responsibility tactics, resettlement and relocation programs, registration requirements, and food controls. In addition to these more common population control techniques, the Communists have developed what has been dubbed the "total societal" approach featuring the simultaneous and coordinated use of social, economic, ideological, and political controls.

The general target for *civic action* is the vast majority of the populace which does not officially participate in the insurgency. There are many methods by which the government may effect civic action programs: it can strengthen the social welfare services to help victims of the underground, expand public health and educational programs, aid agricultural areas, stimulate economic development, and control food prices.

Underground organizations have a number of vulnerabilities which security forces can take advantage of to destroy the movement. The high degree of compartmentalization makes the underground organization vulnerable to infiltration. It is also possible for security forces to play upon the fear of infiltration held by most undergrounders. If the underground can be made to believe that they have been infiltrated, their immediate response is to increase security measures and reduce their operational activities. The reduction in underground activities diminishes the effect of the constant pressure of underground terrorism and agitation.

Since underground communications are organized on a fail-safe basis, once a link is detected it may be placed under constant surveillance in order to trace the other links, perhaps to the underground leadership. In certain situations some underground work, such as finance, training, and supply, may be carried on outside of the country to reduce the possibility of detection and surveillance. Cooperative efforts with other nations or increased border checks can be effective in detecting undergrounders while they are relatively in the open. It is most important in counterinsurgency operations to keep in mind that even when the guerrilla force is defeated the movement is not destroyed until all of the clandestine underground cells have been detected and destroyed.

PART I

ORGANIZATION

INTRODUCTION

The organizational structure of an underground reflects a delicate balance between efficiency and security. While carrying out operations, underground members must be constantly aware of the hostile environment within which they act. The diverse and often conflicting requirements of security and efficiency add complications and anomalies to the underground structure and operations. Many times, in order to achieve one goal, others must be sacrificed.

After many decades of conflict and repeated trial and error, Communist organizational skills and tactics have reached a point of handbook simplicity. Although most of the Communist principles and practices have antecedents in other movements, few organizations have practiced the underground art so widely and so persistently for such an extended period of time.

Although the principles, rudiments, and techniques of political recruitment, organization, and control are elementary and can be found in all societies, their successful application is always impressive.

To fully understand how and why an individual makes certain decisions or takes certain actions, it is essential to understand how he perceives the world around him and to examine the stimuli which impinge upon him within his environment. Whether they are members of family, industrial, or social organizations, persons assume roles which are defined by the nature of the organizations. For this reason knowledge of underground organization is important and prerequisite to the understanding of the behavior of underground members. When an individual joins a subversive organization, the organization becomes a major part of his daily life and alters his patterns of behavior markedly.[1]

If an organization is to achieve its objectives, certain activities, including decision-making and communications, must be carried on. The structuring of these activities provides the context for an individual's behavior and motivation. The roles assumed by the individual, the information he acquires, and the rules, rewards, and punishments imposed upon him by the organization establish the patterns he follows. These structural and organizational determinants of behavior will be briefly reviewed in the first two chapters.

CHAPTER 1

UNDERGROUND ORGANIZATION
WITHIN INSURGENCY

For the purpose of this study, an *insurgent* or *revolutionary movement* is defined as a subversive, illegal attempt by an organized indigenous group outside the established governing structure to weaken, modify, or replace existing governing authority through the protracted use or threatened use of force. An *underground* is defined as those clandestine or covert organizational elements of a subversive or insurgent movement which are attempting to weaken, modify, or replace an existing governing authority.

In its initial stages, when the insurgency is being organized and is necessarily operating in a clandestine manner, the entire organization is considered an *underground*. As the movement develops strength, some elements are militarized and operate overtly. The guerrilla arm is used to combat the military force of the existing government. In this phase the military efforts of the guerrilla units are augmented by the clandestine activities of the underground, which carries on the political war, establishes shadow governments, and supports the military effort. A dual structure of a guerrilla force and a covert underground force appears in most insurgent movements.[a]

INSURGENT ORGANIZATION

Many factors influence the organizational structure of insurgent organizations. The social, economic, and political conditions within the country to a large extent determine who the discontented are, who the participants will be, and what issues and cleavages will appear. Insurgency tends to develop out of internal conflict. Usually the participants do not have access to governmental authority and force, and through protracted conflict attempt to win the support of the people and establish shadow governments.

Terrain and environmental factors also affect organization. Although an underground can function in almost any environment, guerrilla forces are seldom found in harsh climates or highly populated areas. If the leaders of the movement are also members of other organizations, they tend to work within those former organizations and to attract members from them to the underground. Consequently, the character of the former organizations tends to influence the form and character of an underground.

Sanctuary is vital to the existence of an insurgent organization. Neighboring countries or relatively inaccessible rural areas within the country

---

[a] In Malaya (1948–1960), there was an overt armed force, the Malayan Races Liberation Army (MRLA), and an underground force, the Min Yuen; in Yugoslavia (1941–1945), the National Liberation Army and the National Liberation Committees; in Algeria (1954–1962), the guerrilla force, the FLN, and the political force, the ALN; in Vietnam, the Viet Cong's guerrilla arm, the National Liberation Army, and the underground arm, the National Liberation Front.

must offer the insurgents a base area to train cadre and experiment with political appeals and insurgent organization.

External support, primarily psychological but also material, is required if the movement is to survive. International relations have considerable effect on the outcome of internal conflicts.

The form of the underground organization is determined in many respects by the types of people who originate the movement and the environment within which they must operate. If the organizers are primarily military men, the organizational structure usually takes on many of the features of a conventional military organization. If the organizers are politicians, the political role and political aspects of underground activities will be stressed.

Insurgent organizations by necessity operate on both political and military fronts. Not only must they neutralize or destroy the government's military force, they must also win the support of the people and control the people through shadow governments.

The insurgent military force is usually crude and begins with small-unit guerrilla action. If the conflict runs its full course, a regular mobile force, supported by other paramilitary forces, evolves. On the political front, an underground is formed to subvert existing governmental support and organize support for all the insurgents. The underground works through mass organizations and front groups of existing nonpolitical organizational structures and eventually establishes control of people through shadow governments. The underground supports the guerrilla and military front by providing supplies, intelligence, and paramilitary support.

Many times the duties and activities of guerrillas and underground overlap and it is difficult to distinguish between the two organizations. However, several distinctions can be made. Guerrillas have responsible unit commanders, and live and operate outside of the control and surveillance of government forces. Underground members usually live within the control and surveillance of government forces. Their activities may be either legal or illegal, but their goals are illegal within the system and they try to conceal their organization and the identity of their members from the governing authority. All of the civilian organizations associated with an insurgency are defined as underground.

Although the apparent goals of an insurgent organization are well publicized, the true goals may not always be known. For example, in a Communist-inspired insurgency, the Communist Party infiltrates the insurgent organization and creates a clandestine, covert parallel hierarchy within it. In a Communist-dominated insurgency, the underground includes both the civilian organization and the Communist clandestine, covert organization.

MILITARY COMPONENT

The military elements initially employ guerrilla tactics, usually developing a mobile main force later. The regular main force is usually organized along conventional military lines into sections, platoons, companies, battalions, and even regiments. These units operate in the countryside, moving from region to region. The main force is generally supported by paramilitary or guerrilla forces at the regional or local level.

The regional troops, the second element, are assigned responsibility for an area comparable to a province or a state. They move about conducting raids, ambushes, and attacks against government troops. They seek refuge and supplies from local villages at night.

The third element, a local militia, operates from a village and is generally composed of village residents. The members of these units live in their usual way by day and go out on raids only at night.

For definitional purposes, those elements that operate openly, are organized along conventional military lines, and use conventional tactics will be considered the mobile main force. Those overt elements that operate on a full-time basis and use guerrilla tactics will be referred to as paramilitary or guerrilla.

UNDERGROUND COMPONENT

Function

The underground arm of the insurgent movement is usually a hierarchical structure, rising from a base of cells, through branches, districts, states, or provinces to national headquarters. The members may be described as being of three types, depending on their degree of commitment. The leadership cadre is the hard core of the organization and consists of persons who devote full time to the cause. The regular workers continue their ordinary roles in society, but are available to perform organizational duties and attend meetings on a regular basis. The auxiliary, or part-time workers, are available to perform only particular tasks or special assignments.

Another large group is important to the underground—the unorganized sympathizers, nonmembers who participate through such activities as passive resistance and mass demonstrations or by withholding aid and assistance to the government.

While the guerrillas and the main force carry out the insurgent military effort, it is the function of members of the underground to infiltrate and subvert government organizations and institutions. Besides playing an offensive role against the government, they have administrative and organizational roles. They recruit and train members, obtain finances and supplies, establish caches for both the underground and the guerrilla forces, conduct terrorist and psychological operations against the government, and try to win the people's support of the movement. In support of the guerrillas, they are charged with collection of intelligence and with carrying out sabotage against military installations. One of their most important roles is to establish shadow governments and control the people.

The Cell

The basic unit of the underground organization is the cell. It usually consists of a cell leader and cell members. The leader assigns work, checks on members, and acts as a liaison with underground committees. A large cell may require assistant cell leaders. Its size usually depends upon its assigned functions, but in dangerous times the cell is kept small to reduce the possibility of compromise. The cell may be compartmentalized in order to protect the underground organization and reduce the vulnerability of its members to capture. Compartmentalization restricts the information any member has about the identity, background, or current residence of any other cell member. He knows individuals only by their aliases and the means by which they can be reached. This follows the underground "fail-safe" principle: if one element in the organization fails, the consequences to the total organization will be minimal. Furthermore, it is a security measure which protects not only the organization but the individuals in the compartmentalized cells.

The degree of compartmentalization depends upon the size of the organization, the popular support given the government's security forces by the populace, and the probability of detection by security forces. If the security forces have neither instituted population control and surveillance, nor tried to infiltrate the underground organization, the degree of compartmentalization is usually small. At the other extreme, if the populace supports the government and willingly informs it about subversive activity, compartmentalization will necessarily be rigid.

Cells may be organized on a geographic basis or on a functional basis within such groups as labor unions, the professions, and women's organizations. Both types of cells often exist simultaneously. The cells may be highly centralized, with orders flowing from a high command throughout the organization; this tends to increase the efficiency of operations. On the other hand, the organization may be highly decentralized, with units in various parts of the country operating autonomously; this reduces its vulnerability.

The structure of underground cells usually reflects a compromise between requirements of organizational efficiency and the need for security. The structure also varies with the phase of insurgent development.

Structure. The *operational cell* is usually composed of a leader and a few cell members operating directly as a unit. They collect money, distribute propaganda, and carry on the necessary political functions of an underground. (See figure 1.)

The *intelligence cell* is unique in that the cell leader seldom comes into direct contact with the members of the cell and the members are rarely in contact with each other. The structure is such that a member who has infiltrated into a government agency, for example, contacts the cell leader through an intermediary such as a mail-drop, cut-out, or courier. The cell leader is in contact with the branch leader through a courier or mail-drop. Characteristic of this cell is the high degree of compartmentalization and use of indirect communication. (See figure 2.)

The *auxiliary cell* is commonly found in front groups or in sympathizers' organizations. It contains an underground cell leader, assistant cell leaders, and members. Members are usually highly involved in the cause of the

OPERATIONAL CELL

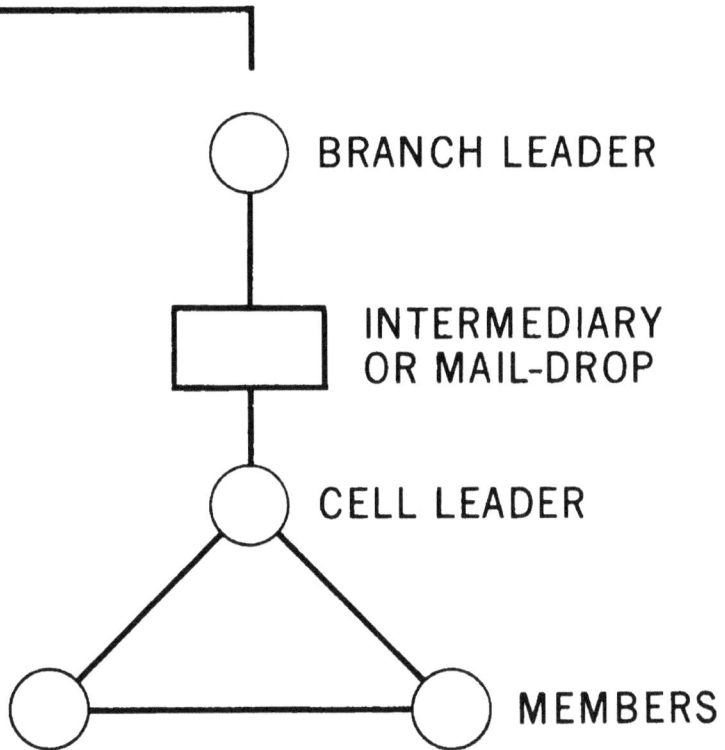

BRANCH LEADER

INTERMEDIARY
OR MAIL-DROP

CELL LEADER

MEMBERS

Figure 1. Operational cell.

underground, but they are either unreliable or untested for routine underground work. The cell leaders identify potential recruits and screen them for the operational underground or intelligence cells. The auxiliary cell differs structurally from the operational cell in that it is larger in size, has an intermediate level of supervision, and has little or no compartmentalization. It is primarily used to handle large influxes of members during an expansion period. (See figure 3.)

Size. Underground operational cells are usually composed of 3 to 8 members.[b] Activities which call for a division of labor require a large cell and a high degree of coordination. The cell may be called upon to serve a specialized function, or it may be asked to work with other cells, each performing part of a complex function in the underground. A big cell with little compartmentalization minimizes the need for formal communications and is thus less vulnerable as far as written records are concerned. However, its vulnerability to capture is greater, because the members know

[b] In the North Korean infiltration into South Korea and in the Communist Party of France during World War II, cells were composed of three members; in the Soviet underground behind the German lines in World War II, in the pre-World War II anti-Nazi campaign, as well as in Cuba, the cells had three to five members; in the Polish underground and in Egypt during Nasser's revolution, clandestine cells were organized into five-man groups; and in Denmark sabotage cells were composed of six members. In Algeria, the FLN's basic unit was a half-cell, with three men; only the leader of each full cell knew the members in each half-cell. Immediately above the cell was a half-group (two cells of seven men each, plus a leader); then a group (two half-groups and a leader); and then subdistricts and districts organized on the same principle.[2]

21

each other and have frequent interaction. If one member is caught and informs, all members will be compromised. In the small compartmentalized cell, the danger that critical underground leaders and cadre will be captured is minimized. On the other hand, it has a greater need for formal communications between units.

INTELLIGENCE CELL

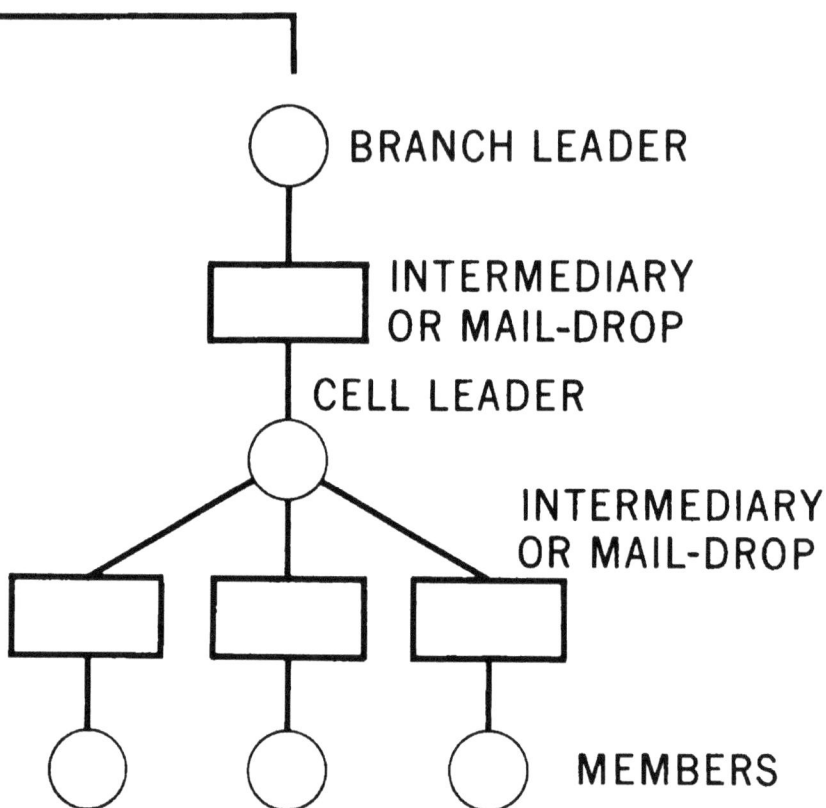

Figure 2. Intelligence cell.

The size of the operational cell also varies according to the phase of development of the organization. Where there is a political party which is legal, the main attempt is to recruit people into the party and then indoctrinate them. In this case the cell may be large. For example, in Germany prior to World War II, the Communist Party cells consisted of as many as 20 members who met twice a week. Each cell was headed by a political leader, an administrative organizer, and an agitprop leader.[3] When it became apparent that the Nazis were gaining control of the country, the Communists prepared to go underground. The cells were reduced in size and compartmentalized to diminish the risk of infiltration by agent provocateurs. Only the leader of each group of five knew the identity and addresses of the other four members of his cell. He alone could contact the higher levels of the party.[4] As a practice, no one person in one group knew the identity or composition of any other group.

AUXILIARY CELL

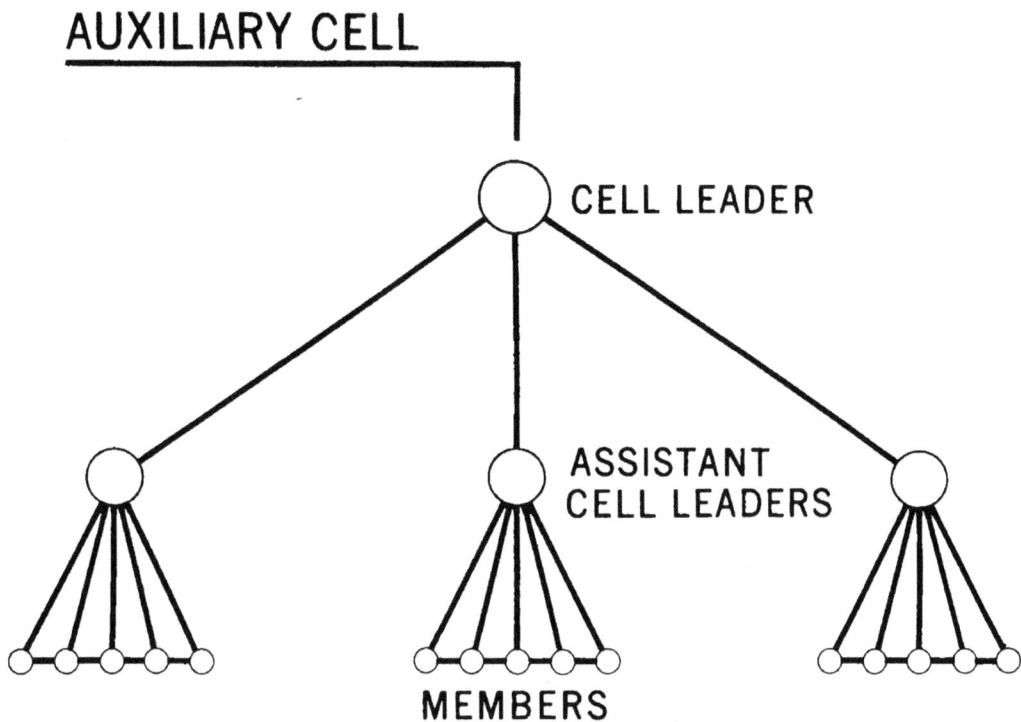

Figure 3. Auxiliary cell.

Similarly, the Communist Party in France before World War II had cells of 15 to 20 and even 30 members. After the party was declared illegal in September 1939, until the armistice in June 1940, cell size was reduced to three men in order to maintain a high degree of security.[5] Later, to increase the party's effectiveness and size, eight-man cells were set up, but between October and December of 1940 the size was reduced to five men. During the German occupation, the party returned to three-man cells in order to insure maximum security.[6] In times of maximum security the three-man cell seems to be the basic unit. But when government security enforcement is relatively loose and there is a need for recruitment, cell membership may be increased to as high as 30.

Critical high-risk cells are usually small, compartmentalized and detached. Intelligence cells are highly compartmentalized and usually maintained at approximately three members.[7] Sabotage units also are usually kept to three-man cells [c] and remain independent of other underground networks.[10] The sabotage units usually work on their own and set up their own communications system.[11] Specialized terror units function in much the same manner and are also kept to three or four members.[12]

Auxiliary cells, such as those in youth organizations, are less compartmentalized and violate many of the rules of clandestine behavior in order to enroll members into the underground organization. These cells act as a

---

[c] In Denmark during World War II, sabotage units were generally 6-member cells that operated autonomously.[8] In Cuba, the cells for sabotage were kept to 3 or 4 members plus a leader.[9]

screening device, testing members before they are accepted into the formal underground organization. d

Number. The number of cells primarily depends upon the density of the population. An underground seeks to disperse its units geographically as well as ethnically. To avoid overconcentration in any one group, organization, or geographic region, which would make surveillance by security forces easier within each area, the underground generally has cells in various blocks, districts, cities, and regions. It infiltrates and also creates cells in existing organizational elements, such as labor, youth groups, and social organizations.

Communist Party members maintain dual-cell membership. The underground member may be part of a cell made up of agents who live within a certain residential area or block: these are called street cells. He may also be a member of a cell at his place of employment: a workshop cell. 14 Dual-cell membership is more or less universal in countries where the Communist Party is legal, and the number of cells a member beldngs to depends on the functions he is to perform.

Parallel Cells. Parallel cells are frequently set up to support a primary cell. (See figure 4.) This is done for several reasons. First, it takes a great deal of time to reestablish cells and if there is to be a continuous flow of information the underground must have a back up cell in case the primary cell is compromised. Secondly, in intelligence, duplicate cells are needed to verify pieces of information and to check the reliability of sources. Parallel cells were set up as a protective measure by the Socialist Party in the anti-Nazi underground.15 communist operations are conducted with as many as four or five independent and parallel intelligence organizations.16 In various front groups parallel cells are used for clandestine support of underground members in the front organization who are seeking positions of authority or responsibility.

Cells in Series. In order to carry out such functions as the manufacture of weapons, supply, escape and evasion, propaganda, and printing of newspapers, a division of labor is required. In the Haganah, clandestine workshops were established to produce small arms. Materials were purchased from regular commercial sources and taken to legal workshops, each of which manufactured components of the weapons. Finally the parts were taken to an assembly plant. The operational cells as well as the operation were organized into a series with management, insuring that the assembly lines were compartmentalized and operated in an efficient manner. Only the underground leader, who kept records of materials, storage, and transportation of the various parts concealed in the company records, was aware of the entire process. Each plant had an intelligence network to act as lookouts.[17]

---

4 In World War *II*, in the anti-Nazi movement, one underground labor youth leader was in charge of 10 subordinates who among them had 90 followers. The members were primarily young students who collected intelligence and gave it to their leaders, who in turn submitted it to the formal underground leader. In Cubs during the anti-Batista movement, a propaganda cell was led by one formal undergroundeader. with 12 subordinates, who in turn controlled 400 members of the propaganda section.10

A similar procedure is used in escape and evasion. The escape network is organized into a chain-like operation where the head of a safe home in the network knows only the next link in the chain and nothing more; an entire escape-and-evasion net is not known to any one individual.

PARALLEL CELLS

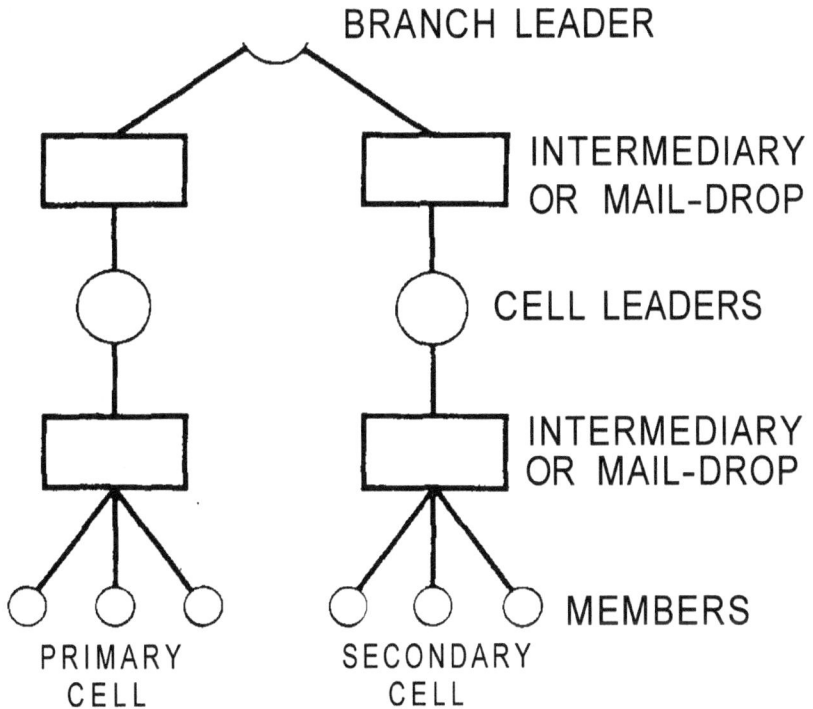

Figure 4. Parallel cells.

In the Belgian underground six cells or sections were connected in a series to produce large-scale newspapers. One cell, composed of reporters, gathered the'information and sent it to a second cell which was composed of editors, who wrote the material. One cell was charged with supply: that is, getting the ink, paper, and lead. Another cell was in charge of administration — keeping books and funds. An additional cell was in charge of the printing; and, finally, through various other cells the newspaper was distributed.18

Often cells are expanded or assembled for a short period to carry out specific, special-mission tasks. In Denmark, small, six-man cells were increased to ten-man sabotage teams in order to carry out large-scale missions. The network eventually included ten teams of ten men each. This was the maximum strength allowed for security considerations.19 In

Greece, terror cells were organized for a particular mission and then dissolved in order to protect the security of the terrorist.[20]

CELLS IN SERIES

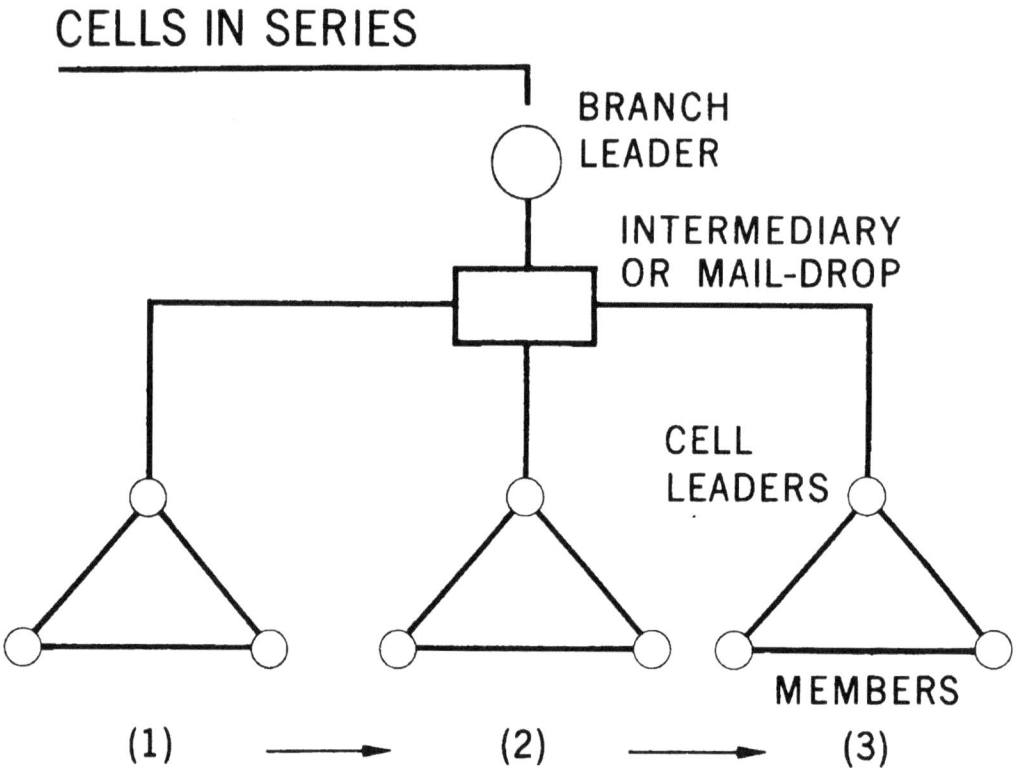

Figure 5. Cells in series.

COMMAND AND CONTROL

Within any organization there is a need for coordination—not simply at single points in time but over a duration of time. The complexities of coordination require some central control. The many activities must be centralized in order to provide subordinate units with services that they cannot provide for themselves. Such functions as strategy, collection of funds, procurement of supplies, and intelligence and security services are usually performed at some central agency.[21]

In conventional organizations, centralization requires a high degree of coordination and coordination in turn requires a great deal of communication. Communication is a serious vulnerability of most underground movements. Frequent meetings, written messages, and records can be used by security forces to identify and destroy the underground organization. There is a great deal of local autonomy with respect to specific actions which require adjustment to local conditions. Tactical decisions are usually made independently by lower-echelon leaders in decentralized commands.[22] Generally, when higher commands issue orders, they communicate them to lower echelons in the form of mission-type orders—orders which say "do whatever is necessary to maximize a certain objective function." [23]

There are two factors that dictate this practice. The first is that the local units probably know the situation better than the central command, and the second is that lower echelons are probably better prepared to make decisions with respect to implementation and time. If a mission or action must be closely directed or if there is a change in strategies and the central command wishes to exercise tight control over the specific units, a liaison representative is usually sent directly to the units to assume control. For routine operations, however, direct control is seldom necessary. One factor which tends to unified action among decentralized units is the long, intensive common training given to the cadre before they depart to assume command of a local unit.

The high degree of decentralization, compartmentalization, mission-type orders and local autonomy of action is primarily a security measure to protect the organization from compromise and is most prominent in the early stages of the movement. However, as the movement expands and the emphasis changes to overt action, main-force units are organized along the lines of conventional command and the underground units become less compartmentalized. A centralized control structure with its direct orders tends to increase the effectiveness and speed of underground and guerrilla action.

There is generally a duplication of command structure with forward and rear elements playing roughly similar roles. In Algeria there was an external command outside the country as well as an internal command within Algeria; in the Philippines there was an internal underground called the "politbureauout," safely located in guerrilla-controlled territory. Similarly, in World War II much of the centralized underground activity was conducted by governments-in-exile and many of them were located in England. The purpose of the external command is to provide alternate command in case the internal one is captured, as well as to permit the necessary command work to take place in a relatively safe location. The internal command is responsible for the coordination of activities within the country.

This dual principle of leadership for security reasons may even extend down to the operational level. In the pre-World War II anti-Nazi underground two types of cells were used. One was composed of members who operated within the country but who were directed by a leader who resided outside the country. This was a security measure to insure continued existence of the cell. A second type of cell was used in which the cadre and cell members both operated within the country. These cells were interconnected and operated through a common directing center. In this second type of cell, organizational security was sacrificed for organizational effectiveness.[24] However, the dual system of operation provided some balance between security and operational effectiveness.

Insurgents organize their areas of responsibility and administrative boundaries so that they do not coincide with those of the security forces.[25] In this manner the insurgents take advantage of the interface problems which exist among government security forces. In most organizations it is easier to send messages upward in the chain of command than it is to send messages laterally to comparable elements. Therefore, in many cases, the crossing of a city limit or a state line takes the insurgents out of one unit's jurisdiction and responsibility and places them under the jurisdiction of another unit of the security forces. The delays and confusion caused by interface problems often provide the underground with the narrow margin of time necessary to escape or go into hiding.

If underground units are centralized or concentrated in one section of the country or segment of the population, as the OAS was in Algiers during the Algerian independence movement, it is relatively easy for security forces to concentrate all their efforts in this area in order to control and destroy the organization. For security reasons it is advantageous to have representatives in every part of the country, at every geographic location, and in every political unit. It is also functionally desirable to use existing organizations, such as unions, military organizations, and political parties, to achieve the purposes of the subversive movement.

In addition to decentralizing and leaving many decisions to lower-echelon units, undergrounds compartmentalize their activities. The result is an organization that is highly individualistic in its operations. This in itself is a security measure, for it makes it extremely difficult for security forces to identify the *modus operandi* of one cell or unit by uncovering or penetrating other cells.

ORGANIZATION AND EVOLUTIONARY DYNAMICS

In the development of an insurgent or revolutionary movement, there are many activities which are not visible to the casual observer. The organization and activities of an insurgent movement have been likened to an iceberg, with the bulk of the organization and its activities lying submerged and only the overt operations of the guerrillas being visible.[26] (See figure 6.)

In a protracted revolution, organizational activities of the underground undergo various changes. Although the phases of change can be identified, they do not necessarily follow a fixed pattern of development. They may overlap and their evolutionary progress may vary in different parts of the country due to local conditions.

In the *clandestine organization phase*, the underground begins by setting up cells, recruiting, training, and testing cadres, infiltrating key industrial labor unions and national organizations, establishing external support, and establishing a base in a safe area. During this phase the organization is small and highly compartmentalized. Cell size is kept small and new cells are added. Operational-type cells are usually maintained with three members each, and intelligence-type cell structures are used for those agents infiltrating key installations and organizations.

In the *psychological offensive phase*, the underground capitalizes upon dissatisfaction and desire for change by creating unrest and disorder and by exploiting tension created by social, economic, and political differences.

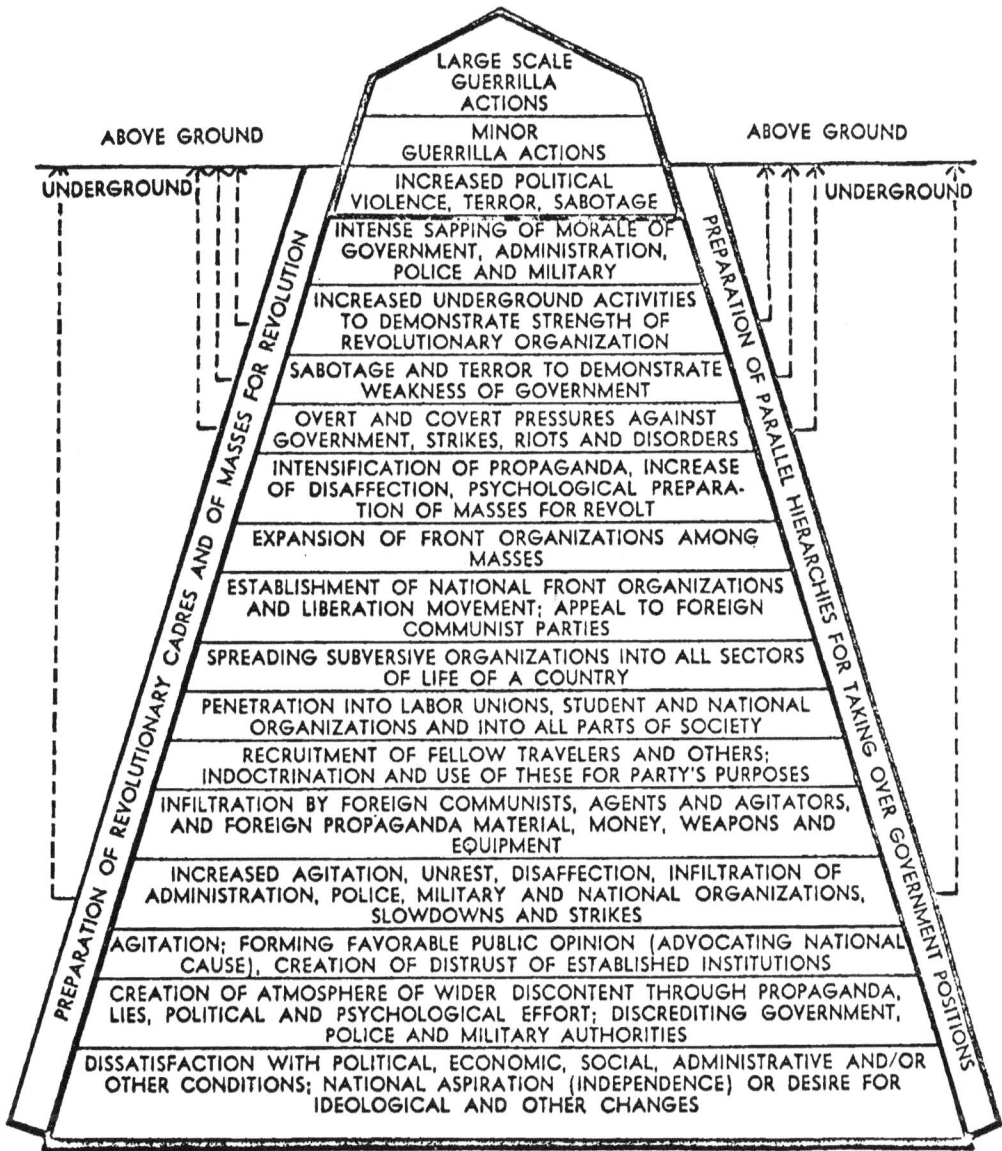

Figure 6. The building of a revolutionary movement.

LARGE SCALE GUERRILLA ACTIONS

ABOVE GROUND | ABOVE GROUND

UNDERGROUND | UNDERGROUND

MINOR GUERRILLA ACTIONS

INCREASED POLITICAL VIOLENCE, TERROR, SABOTAGE

INTENSE SAPPING OF MORALE OF GOVERNMENT, ADMINISTRATION, POLICE AND MILITARY

INCREASED UNDERGROUND ACTIVITIES TO DEMONSTRATE STRENGTH OF REVOLUTIONARY ORGANIZATION

SABOTAGE AND TERROR TO DEMONSTRATE WEAKNESS OF GOVERNMENT

OVERT AND COVERT PRESSURES AGAINST GOVERNMENT, STRIKES, RIOTS AND DISORDERS

INTENSIFICATION OF PROPAGANDA, INCREASE OF DISAFFECTION, PSYCHOLOGICAL PREPARATION OF MASSES FOR REVOLT

EXPANSION OF FRONT ORGANIZATIONS AMONG MASSES

ESTABLISHMENT OF NATIONAL FRONT ORGANIZATIONS AND LIBERATION MOVEMENT; APPEAL TO FOREIGN COMMUNIST PARTIES

SPREADING SUBVERSIVE ORGANIZATIONS INTO ALL SECTORS OF LIFE OF A COUNTRY

PENETRATION INTO LABOR UNIONS, STUDENT AND NATIONAL ORGANIZATIONS AND INTO ALL PARTS OF SOCIETY

RECRUITMENT OF FELLOW TRAVELERS AND OTHERS; INDOCTRINATION AND USE OF THESE FOR PARTY'S PURPOSES

INFILTRATION BY FOREIGN COMMUNISTS, AGENTS AND AGITATORS, AND FOREIGN PROPAGANDA MATERIAL, MONEY, WEAPONS AND EQUIPMENT

INCREASED AGITATION, UNREST, DISAFFECTION, INFILTRATION OF ADMINISTRATION, POLICE, MILITARY AND NATIONAL ORGANIZATIONS, SLOWDOWNS AND STRIKES

AGITATION; FORMING FAVORABLE PUBLIC OPINION (ADVOCATING NATIONAL CAUSE), CREATION OF DISTRUST OF ESTABLISHED INSTITUTIONS

CREATION OF ATMOSPHERE OF WIDER DISCONTENT THROUGH PROPAGANDA, LIES, POLITICAL AND PSYCHOLOGICAL EFFORT; DISCREDITING GOVERNMENT, POLICE AND MILITARY AUTHORITIES

DISSATISFACTION WITH POLITICAL, ECONOMIC, SOCIAL, ADMINISTRATIVE AND/OR OTHER CONDITIONS; NATIONAL ASPIRATION (INDEPENDENCE) OR DESIRE FOR IDEOLOGICAL AND OTHER CHANGES

(left side, vertical) PREPARATION OF REVOLUTIONARY CADRES AND OF MASSES FOR REVOLUTION

(right side, vertical) PREPARATION OF PARALLEL HIERARCHIES FOR TAKING OVER GOVERNMENT POSITIONS

Through strikes, demonstrations, and agitation, a wider atmosphere of discontent is generated. Covert underground agents in mass organizations act in concerted effort with agitators who call for demonstrations and through subversive manipulation turn them into riots. Underground activities are directed at discrediting the police and the military and government authorities. Operational terror cells in many parts of the country operate through the selective use of threats, intimidation, and assassination. The total number of cells in the underground is increased; cells in series are created in order to run underground newspapers, make large agitation efforts, and undertake other such large-scale coordinated activities throughout the country.

29

In its *expansion phase*, after its disruptive activities create unrest and uncertainty, the movement seeks to crystallize public support for a strong organization that will restore order. The emphasis is put on recruiting people through mass organizations and winning popular support for change. Auxiliary cells are created to accommodate new members. Support is built up in front groups and created in other national organizations by covert members. An effort is also made to establish a national political front of many organizations. Trained cadres create new cells and mass organizations. Auxiliary cells are created to handle the influx of new members. Recruiting progresses from being highly selective in the early stages to mass recruitment in the communities and rural areas, and ultimately to drafting young men and women.

The overt activities of the *militarization phase* draw general attention to the insurgent movement. A guerrilla force is formed to harass the government military force. In its tactics the insurgent military force avoids conventional fixed fronts; there is a quick concentration for action and an immediate disengagement and dispersal after fighting.

The guerrilla strategy generally follows the three stages outlined by Mao Tse-tung.[27] The first is called strategic defense. Because the government forces are usually superior, the guerrillas concentrate on harassment, surprise raids, ambushes, and assassinations; they try to force the government troops to extend their supply lines. Since their primary aim is control of people rather than territory, they readily trade territory to preserve the guerrilla force.

The second stage begins when the government forces stop their advance and concentrate on holding territory. As men, arms, and supplies are acquired, the guerrillas attack larger government forces and installations. In this situation, the government is prepared to fight conventional war but the guerrillas are dispersed and capitalize on their speed and mobility. Thus, harassment wears down the government troops while the guerrillas are organizing and building their army. As Mao says, "Our strategy is one against ten and our tactics are ten against one." [28]

The third stage referred to by Mao is the counteroffensive. This begins when the guerrilla army becomes sufficiently well-trained and well-equipped to meet the government forces. The guerrillas seek to create liberated areas; within these areas of control, they build up additional military forces.

The guerrilla force is established only after the leadership has decided that the revolutionary structure is strong enough to support its own army. Underground agents infiltrate towns and villages and begin clandestine recruiting of villages into front groups and local militia. They train and indoctrinate key recruits. Later these groups become feeder organizations for the regional and main-force units.

As the insurgent internal supply arm, the underground purchases supplies, either on the black market or in the legal market through front organizations. They raid warehouses, and set up factories in urban and rural areas. Supply sources outside the country are also tapped through firms that import under noncontraband labels from friendly governments. Caches are maintained throughout the countryside.

The underground provides transportation to move supplies, concealing the load or otherwise discouraging the authorities from making an inspection. As part of the transportation system, storage facilities are provided in houses, central locations, and remote areas.

External sources, such as foreign governments or fraternal societies, are tapped for funds. Internally, loans are obtained from wealthy sympathizers. Other techniques used to raise funds include selling items from door-to-door, robbing wealthy individuals and business firms, coercing people into making contributions, levying taxes in controlled areas, and counterfeiting.

National organizations are subverted by underground members who join the organization and represent themselves as dedicated, loyal members worthy of leadership positions. With the aid of underground cells among rank-and-file members of the organization and a system of rewards, bribes, and coercive techniques, the underground obtains control of many social and political organizations.

The underground forms front groups when it is unable to infiltrate existing organizations. These front groups espouse some worthy cause that will enlist the support of respectable members of the community, but the underground members keep the leadership in their own hands.

The underground communicates propaganda messages by radio, newspapers, pamphlets, word-of-mouth, and slogans and symbols printed on walls. Agitators operate covertly trying to crystallize sentiment for the insurgents. Armed propaganda units go from village to village lecturing on the ways of the organization. Demonstrations are used to show dissatisfaction with the government and commitment to the insurgents. Another technique is to encourage the populace to use passive resistance. By capitalizing on longstanding antagonisms and resistances, the underground attempts to get neutral groups involved in demonstrations. The demonstrators are then moved toward violence as underground agitators create events which lead security forces to take action against the crowd. Through a precipitating event such as an assassination and through the use of agitators within the crowd, subversive agents convert civil demonstrations into riots and violence.

The underground uses terrorism not only to instill fear but to draw attention to the movement and to demonstrate in a dramatic way the strength and seriousness of its operation. A small strong-arm unit, such as most undergrounds maintain to protect their members, may also be used against informers and people who cooperate with the enemy. Because terror is a state of mind, the underground must carefully assess the reactions that follow the use of it.

In selective sabotage the underground attempts to incapacitate installations that cannot easily be replaced or repaired in time to meet the government's crucial needs. Special attention is directed at tactical targets, such as bridges. Sabotage acts are also undertaken to encourage the populace to engage in general acts of destruction. This general sabotage is carried out with such simple devices as Molotov cocktails, tin-can grenades, and devices to cause fire or damage to small items of equipment.

The underground infiltrates agents into government, military, and police organizations and establishes an intelligence organization. Agents living in villages and towns also provide the guerrilla forces with tactical intelligence and local movements of the government forces.

The underground establishes escape-and-evasion operations. Egress routes that direct persons away from lines of battle are set up and fugitives are hidden in secret lodgings, in remote areas, or with guerrilla units.

Finally, there is the *consolidation phase*. While military operations are under way, the insurgent underground continues its political actions. One of the most important functions of the underground is the creation of shadow governments. Initially, infiltrated agents establish covert cells within a village or city. Next, small front organizations are created. Through "persuasion," or with the aid of guerrilla forces, "elections" are held and liberation committees selected on which underground members as well as local villagers are represented. Schools, courts, and other institutions which influence the minds and actions of men are brought under the control of this shadow government. The people within the villages are brought into mass organizations for indoctrination and control over their actions. Undergrounds do not rely on goodwill alone. When in control of an area, they occasionally resort to the elimination of all opposition, and the establishment of covert surveillance systems within the new mass organizations and the civil government. Village by village, the underground takes over and finally governmental support is eroded and an entire area is controlled by the insurgents.

ORGANIZATIONAL INFLUENCES UPON MOTIVATION AND BEHAVIOR

The character as well as the structure of the underground is influenced by the background of the persons who organized it. It will reflect the military, political, or organizational backgrounds of its organizers. The membership in time will be affected by the predominant characteristics of the movement. The leaders of the movement tend to work within former organizations to attract members to the underground and consequently the character of such organizations influences the form and character of the underground organization.

The discipline and sanctions imposed upon members are usually a function of the effectiveness of the security forces. If the security forces are highly effective, the underground tends to be very secretive and disciplined, with severe sanctions for any deviations from the rules of the organization.

Constraints upon what an individual can or cannot do are implicit in organizational membership. Rules for decision-making and communications prescribe certain forms of behavior which members must follow. In addition, organizational rewards and punishments offer new motives and incentives, specifically influencing the member's daily activities and how he performs them.

The structure of an organization will, in itself, influence an individual's behavior. In guerrilla organizations, for example, behavior is conditioned by the kind of unit in which the individual is involved. Mobile main forces are usually large, well-disciplined units, requiring conventional military behavior. Regional forces are made up of smaller units composed of friends and neighbors within a village; operations are only on a part-time basis and discipline is less rigid. In the underground structure, an individual's behavior is affected by the kind of cell to which he belongs. Members of auxiliary cells work intimately with a large number of people; a member of an operational cell comes in close contact with only two or three other members; and a member of an intelligence cell never comes directly in contact with other members of the underground. The type of organizational unit in which an underground member finds himself also determines whether he works individually, as a member of a small group, or as part of a large military unit, what sort of discipline is exercised, and finally, whether he works at home with long-time friends and relatives or away from home with new-found friends or strangers.

The nature of the organizational command-control structure also tends to influence an individual's motivation and behavior. He may follow a strict organizational pattern of behavior or be free to take independent action depending on whether the organization is highly centralized or decentralized. The type of command order, a direct or general group order, will affect an individual's reaction and subsequent behavior. The frequency of command communication determines the extent of individual guidance and control. Behavior is also affected by whether the communications are direct or clandestine through mail-drops or intermediaries.

An individual's tasks and responsibilities influence his motivation. For example, a cadre member, because of his responsible position and power, is likely to be more willing to adopt organizational goals and presumably requires less indoctrination and motivational incentive than other members. A guerrilla in a remote redoubt, having relatively little interaction with people outside of the movement, may not have a strong ideological sense of commitment, but an underground member involved in agitation and propaganda among the masses may find himself believing the propaganda he daily dispenses. Similarly, an underground intelligence cell member who is required to assume a progovernment facade, in order to protect himself from discovery, is greatly influenced in his mode of behavior by the facade.

An underground may also require certain patterns of behavior in order to create a favorable image. Members frequently are prohibited from taking anything from the people without paying for it; there are usually strict rules regarding sex relations among underground members; undergrounders may be directed to befriend certain segments of the population in order to influence them to support the movement.

The phase of insurgent development affects the organizational structure of an underground and, in turn, shapes the behavior of underground members. During the clandestine phase of development, for instance, members refrain from doing anything which draws attention to themselves or to the organization. However, during the psychological offensive and expansion or militarization phases, members adopt a more overt role and attempt to draw the popular attention avoided earlier. Finally, in the consolidation phase, the underground member assumes the role of just and fair administrator in establishing a shadow government.

In short, organizational goals, structure, command and control, and phases of insurgent development all, in turn, help shape an individual's goals, environment, behavior, and motivation. Many of the points discussed briefly above will be dealt with in more detail in later chapters.

FOOTNOTES

[1] James G. March and Herbert A. Simon, *Organizations* (New York: John Wiley and Sons, 1961), pp. 2–4.

[2] For details on the cell size in Korea, see Fred H. Barton, *North Korean Propaganda to South Koreans (Civilians and Military)*, Technical Memorandum ORO–T–10 (EUSAK) (Chevy Chase, Md.: Operations Research Office, 1 February 1951), pp. 151–57; for Denmark, Lt. Jens Lillelund, "The Sabotage in Denmark," *Denmark During the German Occupation*, ed. Borge Outze (Copenhagen: The Scandinavian Publishing Co., 1946), p. 52; for Poland, T. Bor-Komorowski, *The Secret Army* (London: Victor Gollancz, Ltd., 1950), pp. 22–25; for Egypt and Cuba, Paul A. Jureidini, *et al.*, *Casebook on Insurgency and Revolutionary Warfare: 23 Summary Accounts* (Washington, D.C.: Special Operations Research Office, 1962), pp. 364–66 and p. 181, respectively; for France, A. Rossi, *A Communist Party in Action* (New Haven, Conn.: Yale University Press, 1949), pp. 159 and 163; for the Soviet Union during World War II, Otto Heilbrunn, *The Soviet Secret Services* (New York: Praeger, 1956), p. 62; for the anti-Nazi underground, Hans J. Reichhardt, "New Beginnings: A Contribution to the History of the Resistance of the Labor Movement Against National Socialism" (unpublished mimeographed manuscript, circa 1961); and for the FLN in Algeria, Roger Trinquier, *Modern Warfare* (New York: Praeger, 1964), p. 11, and Brian Crozier, *The Rebels* (London: Chatto and Windus, 1960), p. 137.

[3] Arthur Koestler, "The Initiates," *The God That Failed*, ed. Richard Crossman (New York: Harper and Brothers, 1949), p. 42.

[4] *Ibid.*, p. 51.

[5] Rossi, *Communist Party*, p. 159.

[6] *Ibid.*, pp. 162–63.

[7] Barton, *North Korean Propaganda*, p. 122; George K. Tanham, "The Belgian Underground Movement 1940–1944" (unpublished Ph.D. dissertation, Stanford University, 1951); also Reichhardt, *op. cit.*

[8] Lillelund, "Sabotage," p. 52.

[9] Jureidini, *Casebook*, p. 181.

[10] David J. Dallin, *Soviet Espionage* (New Haven, Conn.: Yale University Press, 1955), p. 129.

[11] Ladislas Farago, *War of Wits* (New York: Funk and Wagnalls, 1954), p. 251.

[12] Col. de Rocquigny, "Urban Terrorism," *Military Review*, trans. in XXXVIII (February 1959), pp. 93–99.

[13] Jureidini, *Casebook*, p. 181.

[14] Koestler, "The Initiates," p. 24.

[15] E. K. Bramstedt, *Dictatorship and Political Police: The Technique of Control by Fear* (New York: Oxford University Press, 1945), p. 196.

[16] Koestler, "The Initiates," p. 26.

[17] Gershon Rivlin, "Some Aspects of Clandestine Arms Production and Arms Smuggling," *Inspection for Disarmament*, ed. Seymour Melman (New York: Columbia University Press, 1958), p. 193.

[18] Tanham, "Belgian Underground," pp. 221–26.

[19] Lillelund, "Sabotage," p. 52.

[20] Andrew R. Molnar, *et al.*, *Undergrounds in Insurgent, Revolutionary and Resistance Warfare* (Washington, D.C.: Special Operations Research Office, 1963), p. 306.

[21] Kenneth J. Arrow, "Control in Large Organizations," *Management Science*, X, No. 3 (April 1964), pp. 387ff.

[22] Mao Tse-tung, "Problems in Guerrilla Warfare," *Chinese Communist Guerrilla Tactics*, ed. Gene Z. Hanrahan (New York: privately published, 1962), pp. 10, 61–62; Luis Taruc, *Born of the People* (New York: International Publishers, 1953), p. 115.

[23] Arrow, "Control," pp. 387ff.

[24] E. K. Bramstedt, *Dictatorship*, pp. 196–97.

[25] Trinquier, *Modern Warfare*, p. 80.

[26] Slavko N. Bjelajac, "Principles of Counterinsurgency," *Orbis*, VIII, No. 3 (Fall 1964), pp. 655–69.

[27] Mao Tse-tung, "On the Protracted War," *Selected Works*, Vol. II (London: Lawrence and Wishart, 1954).

[28] Mao Tse-tung, "Problems of China's Revolutionary War," *Selected Works*, Vol. I, *op. cit.;* Vo Nguyen Giap, *People's War, People's Army* (Washington, D.C.: Government Printing Office, 1962), pp. 98ff.

CHAPTER 2

COMMUNIST ORGANIZATION

Revolutions against indigenous governments and organized resistance to foreign invaders have been common in every era of history. Equally common in revolutionary and resistance warfare has been the use of guerrillas and guerrilla tactics. Although the terms "guerrilla" and "guerrilla warfare" originated in the Spanish resistance to Napoleon's occupation, guerrilla strategy and tactics were at that time well known throughout Europe and Asia and can be traced to much earlier times.[1]

However, with the advent of the Russian Revolution, new and significant refinements were added to the strategy of revolutionary warfare. In his 1902 pamphlet "What Is To Be Done?" V. I. Lenin laid the organizational foundations of modern insurgency. He formulated the notion that if revolutions are to be successful they must be led by small, professional (i.e., Communist) elites. Later in his *Left-Wing Communism: An Infantile Disorder*, written in 1920, he stressed the importance of political infiltration and the use of united fronts to disguise the Communist revolutionaries' purpose. He stressed the importance of creating a covert parallel apparatus with interlocking leadership so that a small highly disciplined elite could secretly direct and control a much larger revolutionary movement, which they could then use to achieve the goals of the elite. Thus originated the Communist-dominated insurgency. Mao Tse-tung formalized the strategy and tactics of a protracted guerrilla war among the rural peasantry as a means of extending international Communism into underdeveloped areas of Asia.

When World War II began, many national groups organized underground resistance movements in Europe and Asia to resist occupation and reestablish legal, indigenous governments. The Communists seized upon this ideal opportunity to lay the groundwork for revolutionary movements. They combined the principles of guerrilla warfare with political penetration and control. Throughout Europe and Asia, Communist resistance movements sought more to gain political control than to carry on resistance warfare against the enemy.

In the aftermath of World War II, the Communists were successful in turning resistance movements into revolutionary movements in such countries as Yugoslavia, Albania, and China. In the postwar years, international Communism sponsored and, in many cases, organized and supported "wars of liberation" in Greece, Malaya, the Philippines, Indochina, Cuba, Vietnam, Laos, and Venezuela.

International support of internal subversion has become a pattern throughout the world and was voiced as a policy of international Communism in Nikita Khrushchev's speech "For New Victories of the World Communist Movement" at the November 1960 Conference of Representatives of Communist and Workers' Parties held in Moscow. He stated that wars of national liberation are inevitable and that Communists must fully support them. He thus established the position of world Communism as supporting worldwide insurgency.

In an article *Long Live the Victory of the Peoples' War* in 1965, Lin Piao, Vice Chairman of the CCP Central Committee, Vice Premier, and Minister of National Defense, elaborated upon Mao Tse-tung's theory of the new democratic revolution and reiterated the theme of support for worldwide wars of national liberation. Mao Tse-tung's earlier theory had emphasized the rural revolutionary base areas and the encirclement of the cities from the countryside. Mao has now extended this principle to the entire globe, conceiving of North America and Western Europe as the cities of the world and Asia, Africa, and Latin America as the rural areas of the world which encircle the cities. He maintains that in the final analysis the whole cause of world revolution hinges on the success of revolution in Asia, Africa, and Latin America, since they have the overwhelming majority of the world's population.

According to Mao's theory, the new democratic revolution has two stages: first, a national revolution and then, a Socialist revolution. He maintains that the first is the necessary preparation for the second. He concludes that Socialist countries should support nationalistic revolutions and that these revolutions should be led by a revolutionary party armed with Marxism-Leninism.

INTERNATIONAL ORGANIZATIONS

Article 12 of the Communist International (Comintern) Statutes called for Communists throughout the world to create secret illegal Communist organizations alongside legal organizations. The covert worker of the "illegal" organization disassociated himself from the Communist Party and its members, conducted himself self-effacingly, and cultivated a harmless appearance. The agent used different cover names in different parts of town and changed his cover address and sites for meetings frequently. All his papers and files were kept separate from those of the party organization. No building was used unless its tenants had been investigated.[2]

In the 1930's, the Comintern was extended to every part of the world. In Western Europe and America, permanent bases were established. In many parts of the Western Hemisphere, seamen's and port workers' international organizations served as reception and reporting centers for agents. Agents could report to one of these groups and receive shelter, money, and further instructions. Agents within the maritime organizations made contacts for international functionaries, agents, and instructors passing through their districts and also provided cover addresses for covert communications. They received and handled international funds for local organizations.

During the period from 1930 to World War II, through its executive Committee (ECCI), the Comintern became a second arm of Soviet foreign policy. The secretaries-general of the Comintern's member parties were reduced from leaders of their respective national organizations to mere regional executives of a single structure, for which policy was made exclusively in Moscow.

After 1943, no formal organization existed for the coordination of the activities of the many national Communist parties. At first, some believed that the Cominform (Communist Information Bureau), with headquarters at Bucharest from 1947 to 1955, carried on the functions of the Comintern, at least with respect to the European Communist parties. However, at no time was this body supplied with the staff and clerical personnel necessary to continue the range of activities in which the Comintern had been involved; it was probably no more than what its name suggested—an agency for the distribution of propaganda. Although the Communist International was dissolved in 1943 to rid the party of the propaganda handicap of being an international subversive movement, some authorities believe that international control still exists as a result of the heavy emphasis on indoctrination and institutional character formation of its cadre.[3]

The Communists also distinguish between legal and illegal organizations for gathering intelligence and espionage. While the party operates openly or through front groups, it also operates through embassies, foreign trade commissions, and news agency personnel. Those agencies that enjoy diplomatic immunity are termed the "legal" apparatus within a country. The term "legal" is used because the members of such agencies have diplomatic immunity and, if arrested for espionage activities, are not jailed but declared *persona non grata* and forced to leave the country.

The illegal apparatus is composed of espionage or intelligence agencies such as the Soviet GRU (Military Intelligence Directorate) and the KGB (Committee for State Security) and their agents and informers. If caught and arrested, members of these units can be legally tried for espionage. The GRU is in charge of military intelligence in foreign countries and the KGB units are responsible for nonmilitary espionage in foreign countries, operating parallel with and often rival units to the GRU.[4]

ORGANIZATION OF NATIONAL COMMUNIST PARTIES

Every few years international conferences to discuss and formulate worldwide Communist policy are held. Between these conferences, the national Communist parties are responsible for adapting and implementing conference decisions within their own countries. In recent years the national parties have tended to align with either the Soviet or Chinese Communist parties.

In 1965 there were over 90 Communist parties, with an estimated 44.5 million membership. Parties in 14 Communist countries accounted for 90 percent of the world membership. The Chinese Communist Party of 18 million members is the largest and the Communist Party of the Soviet Union is second with 12 million members.[5]

The supreme authority in each country is the *national congress,* composed of delegates elected by the various conferences and by the next lower level of the party. The national congress meets every two or three years, when convened by the central committee, and is charged with four major responsibilities: (1) to determine the tactical line for the party on political issues; (2) to revise the official program and make new statutes; (3) to hear and approve the reports of the central committee; and (4) to elect the central committee. From time to time it is called upon to discipline top members of the leadership. In practice, most matters which are considered before the national congress have already been discussed. The central committee prepares and documents questions and problems, which are sent to the various party levels where they are discussed and agreed upon. The national congress usually approves what has already been decided. It also sanctions the decisions of the central committee.[6]

The *national conference* is called into special session by the central committee if urgent political matters arise in the period between party congresses. It is restricted in size to a small number of delegates. It is often used as a substitute for the national congress when the party, to minimize the chance of police detection, wishes to conduct clandestine meetings that can be quickly called and dispersed.

Between meetings of the national congress, the maximum authority of the party rests with the *central committee,* composed of top party leaders and varying in size from party to party. The members must have demonstrated competence in organizational ability. Their functions include carrying out the decisions of the national congress, supervising finances, enforcing programs and statutes, and controlling the party press and propaganda. The central committee sets up a finance commission for fund raising and a central control commission to carry out party discipline and security. Its executive bureau, the *political bureau* (politbureau), of 10 to 12 members, is elected by the central committee and directs party activities between meetings.[7]

A *secretary-general* and two aides are elected by the central committee to carry on the daily operations of the party. This secretariat transmits the decisions of the central committee and the party to the subordinate commands. The secretary-general is the highest ranking elected official and is responsible to the party congress. He makes decisions with the politbureau and is responsible to the central committee. The party presidency, an honorary post, exists in some Communist parties.[8]

There are *executive committees* set up to discuss and resolve problems at the various echelons and then pass them to higher authorities for consideration. These committees supervise ideological instruction, the training of executive committees for finance and control committees, and the elections of delegates to the next higher level of the party. In addition, they are responsible for routing party business and directives through their area of jurisdiction, and executing the decisions of the party. All members of the party belong to a cell and have weekly or biweekly meetings.[9]

Cells may be organized geographically, with all members living within a certain territory, or functionally, with "shop units" organized according to type or place of employment. Often they are set up in both ways.

The major emphasis and fundamental principle of organizational work in the Communist Party is to create cells within nonparty organizations, no matter how small the number of Communist sympathizers within an organization. In trades and factories, professional associations, peasant and front groups, and similar places, the Communists are instructed to organize party members into a small group called a "fraction." The fraction consists of part or all of the members of a cell, selected by the Communists to work within existing legitimate organizations. It is the fraction's responsibility to learn the interests, language, and attitudes of the organization, so that they can effectively communicate and disseminate the party propaganda line. They also identify and investigate individuals who may be sympathetic to the party and organize them. In order to guide the fraction, which may include all members from several different trade unions, a nucleus is organized to work under the direction of the local party committee. The nucleus is the "shop unit" and may consist of as few as three members in any one place of employment. The purpose of the fractions is to disseminate the party line, to attract new members to the party, and to aid in developing a power base for the party.[10]

COMMUNIST PARTIES IN NON-COMMUNIST COUNTRIES

The Communist Party, in countries where it is a legal political body, has two major organizations, one open and one covert. The overt organization functions as an ordinary political party. However, the Communists everywhere organize their party into a system of cells and committees, regardless of the size or strength of the party or the degree of government opposition. Even in the "legal" party, the cellular structure serves to train members in conspiratorial behavior. Cell meetings are often held secretly so that members attending them can learn how to travel to and from them without arousing suspicion. Members are assigned minor intelligence-gathering or sabotage missions which in themselves have little or no practical use but which test and train members in clandestine behavior. A press (either open or clandestine) is usually set up in order to give members experience in writing, printing, and distributing material for the party.

In addition to the open legal party, a highly compartmentalized, clandestine organization is also created. Members of its cells are people who have potential value to the Communist Party in the event of an insurgency or coup d'etat. Individuals recruited from government, vital communications centers, industry, or other organizations that the Communists seek to infiltrate may not be admitted directly to the party itself but may become members of clandestine cells. Thus a network of infiltrators and agents in important positions is in readiness. These cells may remain dormant for many years, being used only to collect selective intelligence, to be activated only in case of an insurgency.[11]

The Communists also organize "front" groups to use people who are sympathetic with causes which the Communist Party promotes but who are either unreliable or would not for personal reasons join the Communist Party. Front groups are organized around currently popular issues. While the party usually controls these organizations, they are kept separate. The Communists also attempt to gain control of governmental agencies through coalitions with other, non-Communist parties.[12]

The Communist cadres are full-time professionals who accept the serious risks of revolutionary leadership, and the formation of such cadres is the basic work of the Communist Party. Lenin believed that only a small, militant organization could bring about revolution. The organization proper must be confined to a small, hard core of dedicated individuals who can be counted on to maintain their own discipline and carry out orders precisely and without questions. The term "cadre,"—a group or body of professionals who train and recruit new units around them—is applied to the small Communist vanguard who are to lead the revolution.[13]

The party seeks to create in its cadres a body of men capable of implementing a dictated strategy with "great ability, skill, and real artistry." But such competence cannot be acquired through theoretical studies alone: the member must be constantly tested in political combat. "Each Party can master the art of political leadership only from its own extensive experience." Out of this crucible comes an *aparatchik* who is more than an adherent of a political doctrine: he is a person totally committed to, and with no life outside of, the party.[14]

The Communist Party is highly selective in its recruiting. The potential member must show through practical work that he understands the party and is prepared to accept its discipline. Membership can be conceived of as a process rather than as a condition. The granting of a party card is not the completion of a period of preparation, whereafter the individual can relax with the assurance of having "passed the test." It is in itself only a halfway house of a process whose end product is total mental commitment.

Normally a substantial portion of those who become members do not complete the process by proceeding on in toward the center of the apparatus. Many withdraw along the way, and the party is prepared for this. Only a candidate who is well along the way toward total commitment is permitted to learn the inner workings of the party. He goes through extensive indoctrination courses in discussion groups and party schools. He must participate in the organization of rank-and-file members. These activities are designed to guarantee his total involvement and commitment to the functioning of the party. He must be willing not only to perform legal political activities but to carry out illegal work when required.[15]

The Communist Party operates on the rule of *democratic centralism*. Within the hierarchy of party organizations and committees, each lower body selects a representative to serve on a party committee; this committee in turn selects another representative to serve on a higher committee in the hierarchy.[16] The principle of democratic centralism is followed throughout Communist organizations. Unit committees are elected by the membership or the delegates of the party organization. Each committee must report regularly on the activity of the party organization and must give an account of its work. These committees are responsible for carrying out the decisions of the higher party committees. All decisions of the higher committees are binding upon the lower body members.

In theory, each proposal is discussed at the lower levels of the party, and each committee member presents the opinion of the lower body to the next higher body until a decision is made at the central committee level. Once a decision is made, the entire party must carry it out. In general practice, a decision is determined at the central committee level and, although the lower echelons discuss it, the members are well aware that they must ultimately concur in it.[17]

Elections for committee members and their secretaries must be "approved" by the committee at the next higher level. This enables the leadership to exercise strict control over subordinates and to suppress any opposition from the outset.[18]

This disposition of authority follows the party principle of "reverse representation" at all levels. The "elected" or designated leader of any organizational element of the party, regardless of the level at which he operates, represents among his associates the authority of the next highest party body. He is not the spokesman for his subordinates in high party councils, but rather the latter's liaison with lower levels.

Institutionalized *criticism* or *self-criticism* serves two essential purposes in the Communist organization: (1) it increases the efficiency of the party by subjecting its operations to constant review and revision; (2) it creates a norm of behavior in members and helps secure absolute commitment and dedication to the party.

The actual activities of criticism and self-criticism sessions consist of conferences, discussions, and meetings within the party in which attempts are made to determine and correct any weaknesses in the work of the party or party members. Criticism is practiced on all occasions and is an integral part of Communist life. Theoretically, all decisions and basic policies of the party are open to criticism and discussion in these sessions. But in actual practice, criticisms must never contradict the essential party line and are directed only to improving the practice and implementation of existing revolutionary theory. A member is expected to analyze mistakes and shortcomings of the party operations only. Unless his criticism is constructive—that is, offers a concrete proposal for improvement in work or a method for correcting mistakes—it is not accepted and the individual making the criticism may find himself under attack. No criticism may be made of the central leadership, and no organized expressions of criticism or dissent are tolerated.[19]

Criticism and self-criticism sessions are designed to develop absolute commitment and ideological dedication among members, so that party orders are implemented, not mechanically but creatively. They attempt to make the individual member think in terms of a vanguard and how better to advance the current line and more effectively carry out revolutionary work. Members are compelled to report errors, mistakes, or weaknesses displayed by all party members no matter how small or trivial; they may also state and restate any change in policy.

The sessions establish and reinforce complete ideological unity among the membership. Each individual must conform to the party line. The meetings act as constant reminders of the need to raise their goals, increase their activity, and execute orders faithfully.[20]

Every party member knows that if he does not make every effort to contribute seriously to criticism of his fellows, then in the subsequent comprehensive dissection of his own conduct, he will be obliged to confess this guilt. He also knows that participating fully in the identification of others' failings will not help him to escape his own eventual subjection to the same process. The thorough analysis of his conduct can proceed into the smallest details of his life, both private and public, both intimate and generally known. He must clearly acknowledge his faults before the group and promise to improve. He understands that an inadequate response in his own session can lead to reduction of rank or even to expulsion from the party. Thus the sessions instill in each member a need to demonstrate to his associates his unqualified responsiveness to the wishes of authority so that he can avoid undue attention by his cohorts and escape excessive criticism when his turn comes. In this fashion the Communist Party maintains a built-in, permanent uncertainty and apprehensiveness among the rank-and-file, and can be certain of obedience from below.[21]

Another characteristic of Communist organization is *collective leadership*. Executive and administrative decisions must be agreed upon by the majority of the officers at a given level of the party. Collective leadership, however, is an exception and practiced only during interparty conferences. In practice, the party functions in a highly centralized manner with authority and command decisions flowing from the top to the lower echelons.[22]

Changes in leadership within the Communist Party are not frequent; "elections" become the equivalent of promotions. The leadership submits candidates and issues to the membership. In order to legitimize this authority, members are compelled to discuss these matters and overtly agree and vote on them.

Issues submitted to vote are not appeals to the membership for action, as they commonly are in unmobilized and unstructured groups, but are instead specific orders and plans for future work. The member's attention is focused not on acceptance but upon what is to be done next. Emphasis is placed upon uniformity of thought and the ultimate authority of the leadership.[23]

COMMUNIST USE OF MASS ORGANIZATIONS

Large groups which the Communists strive to infiltrate are called "mass organizations." Communist Party theory holds that a small group of highly disciplined individuals, operating through mass organizations, can rally the support required to win a revolution.[24]

V. I. Lenin recognized the vulnerability of mass organizations to infiltration and manipulation. In turn, Joseph Stalin argued that Communist Party members must avoid the concept that efforts to build up the party should be directed solely to recruiting new members. Instead, he suggested that the Communists systematically use mass organizations as "transmission belts" to the broad masses of nonparty workers. By working through mass organizations, Communist Party workers can reach and influence many thousands of workers "not yet prepared for Party membership." Through "these organizations, led by well-functioning fractions, the Party must necessarily find its best training and recruiting ground. [Mass organizations] are the medium through which the Party . . . guides and directs the workers in their struggles and . . . keeps itself informed on the mood of the masses, the correctness of the Party slogans, etc." [25]

The Communists feel that the simple creation of disorder is not sufficient to bring power into the hands of the elite. They attempt to separate the existing leadership from the institutions and support on which it rests. While disrupting the government, Communists seek to construct new instruments of power. They build their own covert controls within existing organizations or form new organizations which they can control. They try to subvert institutional loyalties and create new allegiances within mass organizations at the community level. They undermine old forms of authority and create new ones, corrupting the authority upon which institutional foundations are built.[26]

OBJECTIVES IN CONTROLLING MASS ORGANIZATIONS

The objectives of infiltrating mass organizations are (1) to neutralize existing agencies which support the government; (2) to justify and legitimize causes which can be exploited by the subversives; and (3) to mobilize mass support.[27]

By penetrating organizations and institutions within the society, the Communists avoid being isolated, and are in a position to neutralize competitors and monopolize mass support. The strategy of neutralization has played a large role in the relationship of the Communist Party to Socialist and other leftwing organizations. They try to infiltrate these groups and through disruptive practices in the organization neutralize their effectiveness and put the leadership in disrepute. Where communism has no popular following in its own right, Communists have sought to mobilize popular sentiment around legitimate issues and causes and so indirectly gain legitimacy for their movement. Another major objective is to mobilize those large segments of society who are not members of groups into formal organizations.

MASS ORGANIZATIONS IN COMMUNIST INSURGENCY

The use of mass organizations in an insurgency can be illustrated by several cases. The Malayan Communist Party (MCP), early in its history, set about organizing a number of front groups, including the Proletarian

Art League, Youth Corps, Racial Emancipation League, and General Labor Unions (GLU). The labor front was perhaps the most important. Using the demand for higher wages to match the rise in rubber and tea prices as a basis, the union movement organized a number of strikes and collective actions.[28] The principle of organizing labor for collective action was new to Malaya in the thirties, and the Communists' efforts to develop labor unions were to pay off in the insurgency after World War II.

After the war, the MCP set up additional organizations. It organized a General Trade Union and a Youth League to attract Chinese students. Once the insurrection was underway, a Cultivation Corps, an Anti-British Alliance Society, a Students' Union, Women's Union, etc., as well as less overtly political organizations, such as youth and sporting groups, were organized. With employment hard to find, it was often necessary for a man to join a Communist union in order to get a job. The MCP also established its own schools and clubs, so that it could approach the Chinese community to conduct political discussions and disseminate party literature.[29]

During the insurgency in Greece, the Communists organized and controlled many front groups, such as the Seamen's Partisan Committee, the Communist Organization for Greek Macedonia, the Democratic Women's Organization of Greece. In the rural areas the Communist Party operated through the Greek Agrarian Party (AKE) and the United All Greece Youth Organization (EPON).[30]

In the Philippines, Communist Party officials spent much time before the war engaging in labor activities in Manila and other parts of Luzon. The printer's union was influenced by Mariana Balgos, and the League of the Sons of Labor was headed by Crisanto Evangelista, both noted Communist leaders. The League of Poor Laborers, the predecessor of the Confederation of Peasants, was among the mass support organizations which provided the base of support for the insurgency between 1946 and 1954. Most of the members of the Communist politbureau in Manila were officers in the unions affiliated with the Congress of Labor Organizations (CLO).[31]

METHODS OF CONTROLLING MASS ORGANIZATIONS

The fundamental aims of most mass organizations are those for which they are organized: labor unions, for example, are organized to improve the lot of the worker. But organizations formed primarily as pressure groups can be used for other purposes. To the Communists, they represent a chance to manipulate the social and political ideas and attitudes of the members.

Most voluntary, large-scale organizations are composed of a leadership (a small corps of individuals who represent the administration), a few faithful, full-time followers, and a large group of dues-paying members. The followers usually leave the operations and decisions to the leadership. Members may or may not agree with the leaders on all decisions and actions.[32] Members who are willing to work and accept responsibility are usually given the opportunity to do so and, indeed, such willingness leads to a gradual promotion to leadership responsibilities.

When planning a takeover, the Communists first try to gain influence in the organization's membership office in order to control recruitment and to infiltrate Communist members. Once in, Communists are instructed to volunteer for all positions and for all work in the organization. They are instructed to be the first to arrive at and the last to leave meetings. They are taught how to harass non-Communist speakers and, through the tactics of attrition-through-tedium, win votes and offices within the organization.[33] They seek the leadership of political and education committees, and use these offices to identify people in the organization who might be sympathetic and those who are avowedly anti-Communist. Editorship of the organization's newspapers provides opportunities for expressing subversive ideas and gives access to printing materials which may be used to establish their own distribution routes. Once they have organized cells or fractions within the organization, they caucus and plan their organizational moves in advance.[34]

The Communist seeks leadership positions and represents himself as dedicated and loyal to the organization, taking the initiative in planning activities and volunteering for any job, no matter how time-consuming or unpleasant. He is instructed to avoid the appearance of any subversive activity. Although his candidacy is supported by cell members in the rank-and-file, close ties between the candidates and the cell collaborators are hidden from the general membership so that the candidates' support appears spontaneous and unsolicited. Usually the most vocal members at a meeting pass resolutions and manipulate the apathetic majority. Therefore, a small, articulate group can readily influence the direction of the organization and eventually gain control.

In such organizations as labor unions, systems of rewards and punishments can be utilized to maintain the obedience of members. If a man is dropped from a union he may not be able to get employment. On the other hand, if the union leader improves the lot of the union members, they will more willingly go along with more purely political actions and obey strike calls. In addition, goon squads may be used to "persuade" uncooperative members. Having instruments of persuasion and coercion, the leadership can gain compliance of a majority of members. Most members will comply with a strike decision, since higher wages may benefit them and failure to comply will only lead to punishment, loss of membership, or worse.[35] In using front organizations, the Communists attempt to develop and maintain the loyalties of people who otherwise could not be persuaded to enter the Communist Party or who, even if willing, would not be sufficiently reliable. They also are able to mobilize many who are indifferent or even opposed to Communist ideology—uniting them instead behind causes such as "nationalism," "liberation," "pacifism," or other popular social issues within a particular society. The organization also attempts to gain support of those elements within the community, such as religious and fraternal organizations, that command the respect and loyalty of the workers.[36]

The cell attempts to evaluate the power structure of the group which it is trying to infiltrate. In professional groups such as industrialists, lawyers, or university presidents, the Communists seek to control executive staff functions since this is where the power resides in such organizations. They look upon the facade of distinguished citizens on the board of directors as an asset to the organizational infiltration. Hence, they do not seek the prestige positions but instead positions of control which affect the day-to-day operations of the organization.[37]

In the Malayan Communist insurgency, for example, the MCP maintained its influence within the General Labor Union through three separate control systems. The first system was made up of a president or secretary and two or three full-time organizers, who were part of the open membership of the labor union. Although they were Party members, they avoided any connection with meetings or activities that might identify them with the party. They reported to and took orders from the GLU. They were told to operate within the law and to give the impression that their primary interest was the advancement and concern of trade unionism.

The second system of control was exercised through underground party members, who held no official office and were members of the open rank-and-file. They were activists who recruited new members for the union and for the MCP. They served to simulate grassroots sentiment for policies favored by the party, enabling the leader to avoid the appearance of dictating to the union. This group also reported on the financial status of each member and provided information to the party on membership attitudes; the party then based organization policy on these reports. The underground members reported to the section of the party responsible for trade unionism, which was separate from the regular party. These members were more trusted than the leaders, who were considered expendable if discovered.

The third control system consisted of the regular party members who formed a fraction within the union membership. They held no official posts. They reported their activities to and received orders from the regular party.[38]

UNITED FRONT ACTIVITIES

The term "front" has been used in three different ways in Communist political warfare. Commonly, it refers to political activities carried out behind the facade of an apparently non-Communist organization. The front has been used to gain control of peripheral leftist groups. It has also been used to gain access to wide segments of society having no ideological commitment to communism or Marxism. In a "united front" operation, the Communists seek to consolidate and unite forces of discontent against the government. The groups in the united front need not agree with the objectives or goals of the Communist Party. However, the party does offer its support. In this manner the Communists maintain organizational integrity while becoming associated with other, legitimate organizations.[39]

The Communists have utilized the technique of the united front in most of their insurgencies. They form alliances with other groups, offering them the organizational support of the Communist Party in return for a united front against some issue. The rank-and-file of most organizations are more than willing to accept anyone who professes to share their views. Many organizations have assumed that the Communists would enter into cooperative ventures without subverting the organization and that their cause would benefit from the additional strength of the Communist Party.[40]

Lenin's formula was to go where the masses are located, vie for leadership positions or neutralize the existing leadership, and gain access to the rank-and-file. By drawing a number of legitimate groups into a united front, the Communists can gain the prestige of speaking for a large and diverse group of people. Once in the front, they seek to discredit the leaders of the other organizations so as to gain control of their followings.[41] Usually organizations join a mass front or coalition in order to achieve particular ends; the Communists join for an opportunity to subvert them. The theory is to fill power vacuums and to create new organizations to cope with new problems which are not being effectively handled within the context of existing organizations.

In Venezuela, the creation of a united front was the first major step in initiating an insurgency and is characteristic of most Communist-dominated insurgencies. For example, the Venezuelan Communist Party (Partido Comunista Venezolano—PCV) initially used its legal status to cover its illegal activities. Communists infiltrated the Democratic Action Party (Acción Democrática—AD) and in 1960, under Domingo Alberto Rangel, the leftwing of the party was expelled. The party formed a new group called the Movement of the Revolutionary Left (Movimiento de Izquierda Revolucionario—MIR). In the mid-fifties, both the MIR and PCV were in militant opposition to President Betancourt. In 1961, the MIR used its congressional immunity to carry on terrorism against the Betancourt regime. Finally, in 1962, both the PCV and MIR were ruled illegal by the Supreme Court of Venezuela. After this decision a National Liberation Front (Frente de Liberación Nacional—FLN) was formed to unite all leftwing elements against President Betancourt and initiate an insurgency. The FLN organized the Armed Forces of National Liberation (Fuerzas Armadas de Liberación Nacional—FALN) to conduct urban terror and guerrilla warfare against the government.

In 1962, the Minister of Defense of Venezuela described the Communist plan as it appeared in captured FALN documents. The FALN proposed: (1) agitation against the government; (2) demonstrations, disturbances, strikes, and terrorism; (3) sabotage and guerrilla actions throughout the country; and (4) insurrection culminating in violent takeover of power. The purpose was to create such chaos that the armed forces would take power through strong-arm methods; then the Communists would overthrow the army and gain control of the government. The Communists have gained support in the left wing of the Republican and Democratic Union (Unión República Democrática—URD) which withdrew from the Betancourt coalition in protest against the government's anti-Castro action in 1960, as well as the AD opposition which split from the AD in 1962. Both of these elements supported the terroristic campaign.[42]

In May 1941, members of the Indochinese Communist Party (ICP) formed the Viet Nam Doc Lap Dong Minh Hoi (League for the Independence of Vietnam) or, as they were popularly known, the Vietminh. The organization was a broad coalition of political parties, all of which wished to free Vietnam from French rule. Since many nationalists did not join the coalition between it was Communist controlled, Ho Chi Minh officially dissolved the ICP on November 11, 1945. In May 1946, in still another move to win nationalist support, he announced the establishment of the Mat Tran Lien Hiep Quoc Dan Viet Nam (Vietnamese Popular National Front), a broader front than the Vietminh, whose goal was "independence and democracy."

In 1951, since the front received most of its aid from Communist China and the Communist bloc, the Communists felt that they had sufficient control over the movement that they could reestablish a Communist Party. In addition, if some unforeseen event should occur in which they lost control of the front, they wanted to leave some official representation in the organization. The name, Dang Lao Dong (Workers' Party), was carefully selected. One party document describes the reasons which went into the choice of name.

> It should never be admitted outside Party circles that the Workers' Party is the Communist Party in its overt form for fear of frightening and alienating property owners and weakening national unity. To party members and sympathizers it can be admitted that the Workers' Party is the Communist Party, but to others it should neither be admitted nor denied . . .[43]

In this way they avoided alienating people who for one reason or another could not accept communism, but at the same time won recognition from other Communist parties throughout the world.

Using the same tactics as in the war against the French, in 1962 the Communists organized the National Front for the Liberation of South Vietnam (NFLSV). In order to control the movement, key members of the central committee of the Lao Dong Party went to the South to run the operations. Once they had firm control, a thinly disguised Communist Party (People's Revolutionary Party—PRP) was formed which is ostensibly independent of the North but in fact is an extension of the Lao Dong Workers' Party.[a] Except in the rare cases involving attempted coup d'etat, the creation of a united front has preceded the initiation of Communist insurgency and guerrilla warfare.

---

[a] When Radio Hanoi announced the formation of the PRP, it avoided the word "Communist" and described the party as "representatives of Marxist-Leninists in the South." A captured Viet Cong document which originated in North Vietnam and was sent to a provisional party committee in South Vietnam, states that the formation of the PRP should be explained to party members as a tactical move to rebut accusations about the invasion of the South by the North to permit the NFLSV to recruit new members and win sympathy and support from nonaligned nations. It goes on to say that the independence of the PRP is only apparent and that in reality the party is the Vietnamese Workers' Party united in North and South under President Ho Chi Minh.

In July 1962, when the North Vietnamese signed the international agreements on Laos, a member of the delegation reported to foreign journalists that the list of members of the central committee of the Workers' Party was necessarily incomplete. Some names had been left off in order to protect the identity of men who were directing military operations in South Vietnam.[44]

COMMUNIST INSURGENT ORGANIZATION

UNDERGROUND AND GUERRILLA STRUCTURE

Far-reaching organizational changes are required to convert a peacetime party into a national liberation movement designed to carry on a protracted revoluntary war. While the party apparatus itself remains essentially the same, an additional structure, composed of underground members and guerrillas, is created. Through interlocking positions of leadership within the movement, the party directs the underground and guerrilla organizations and operations.

At the top of the organization is a central committee, a politbureau, and a secretary-general. The secretary-general directs the national committees for military organization, mass organization, education, finance, and intelligence. Below the national level are provincial, district, and local committees and individual cells. There are two parallel national organizations, one civil and one military. (See figure 7.)

The civil organization or national liberation front is usually made up of several political parties and affiliated mass organizations. The front is responsible for mass recruitment and for support for the guerrillas in the form of intelligence, supplies, and safe-homes. It is also responsible for population control and the establishment of a shadow government to provide schools, courts, taxation, and administrative offices. The front has liberation committees in regions, districts, villages, and towns. Within each of them is a parallel covert Communist organizational element.

Three major military forces evolve. The mobile main force, the regional forces, and the local militia. The first is a regular army and the last two are paramilitary. The regular mobile main-force units are organized on the basis of battalions, companies, platoons, and squads. There is a heavy emphasis, however, on political indoctrination. Each unit generally has a military leader, a political leader, and a party member who make tactical and political decisions.

The military organization is set up according to the principles formulated by Mao Tse-tung. Vo Nguyen Giap, in elaborating on Mao Tse-tung, distinguishes three military phases of insurgency: "guerrilla warfare," "mobile warfare," and "entrenched camp warfare." [45] Certain organizational and political developments are prerequisite for transition from one phase to another.

The guerrilla forces carry on a war of harassment until basic political structures can be created. The more proficient of these forces are then singled out to launch the mobile warfare phase. New main-force units are created and organized along conventional lines but remain mobile and use guerrilla paramilitary forces as a protective screen. Although this mobile mainforce assumes a formal structure of battalion strength, it may fall short of conventional military strength and firepower. By contrast, the guerrilla paramilitary forces retain a simpler organizational form, being composed of small units approximately the size of platoons, and

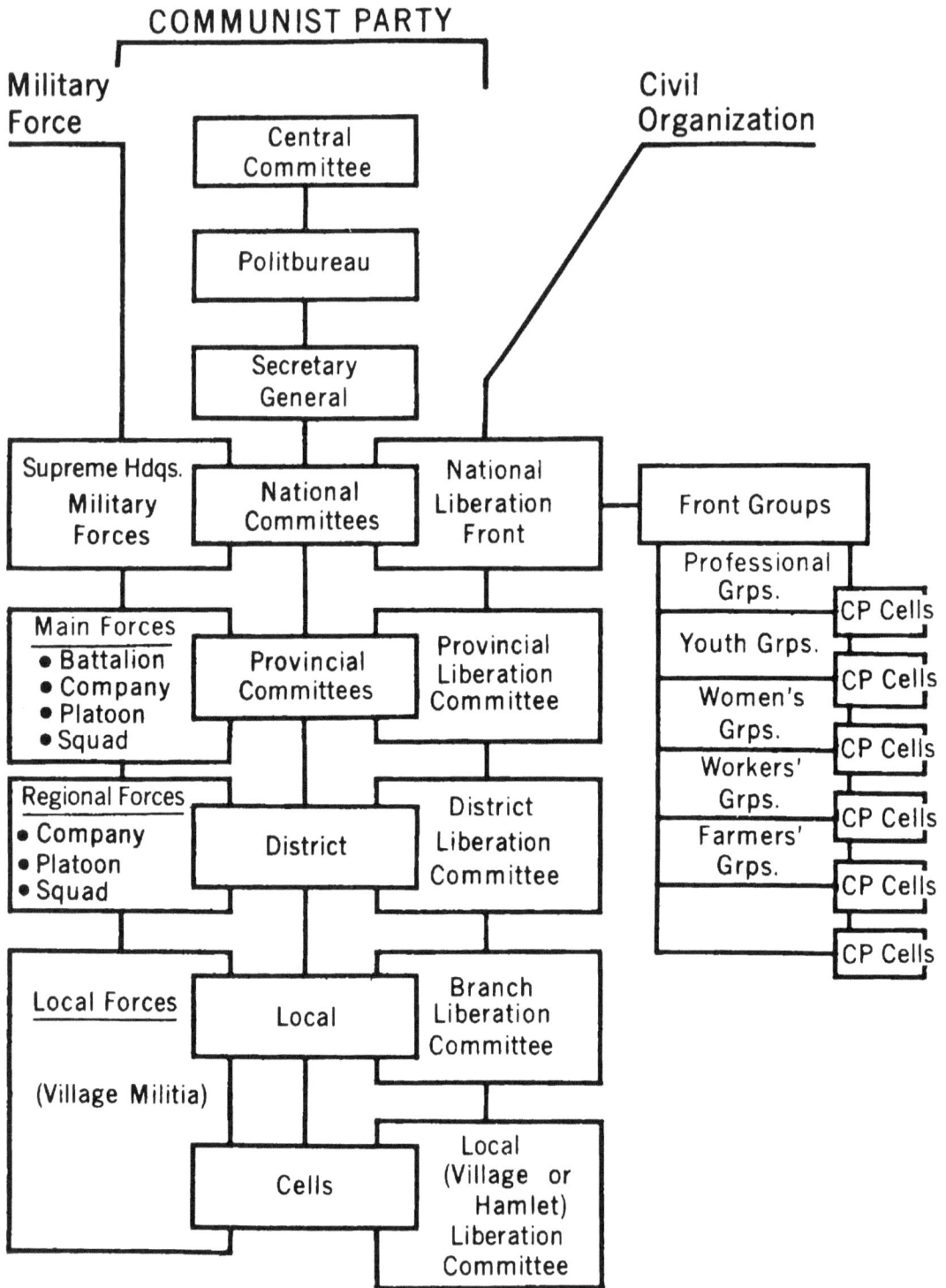

COMMUNIST PARTY

Military Force

Civil Organization

Central Committee

Politbureau

Secretary General

Supreme Hdqs. Military Forces

National Committees

National Liberation Front

Front Groups

Professional Grps.

CP Cells

Main Forces
- Battalion
- Company
- Platoon
- Squad

Provincial Committees

Provincial Liberation Committee

Youth Grps.

CP Cells

Women's Grps.

CP Cells

Regional Forces
- Company
- Platoon
- Squad

District

District Liberation Committee

Workers' Grps.

CP Cells

Farmers' Grps.

CP Cells

Local Forces

Local

Branch Liberation Committee

CP Cells

(Village Militia)

Cells

Local (Village or Hamlet) Liberation Committee

Figure 7.   Communist insurgent organization.

continue their harassment of government security forces. It is the responsibility of local forces to maintain a presence among the civil populace and to harass government forces. Giap maintains that even in the mobile warfare phase a judicious balance must be achieved between these two forces.

The importance of maintaining both a main force and paramilitary formations was also stressed by Mao Tse-tung:

> Considering the revolutionary war as a whole, the operations of the people's guerrillas and those of the main forces of the Red Army complement each other like a man's right arm and left arm; and if we had only the main forces of the Red Army without the people's guerrillas, we would be like a warrior with only one arm.[46]

The militarization phase of the insurgency is simply part of the larger political struggle. As one former high-ranking Communist has remarked, guerrilla war is just a feature of the overall political war.[47] The character of Communist insurgent organization as it is transformed during a militarization phase of insurgency has been illustrated in Malaya, the Philippines, Greece and South Vietnam.

In Malaya, the insurgent movement was made up of the Malayan Communist Party (MCP), the Malayan Races Liberation Army (MRLA), and the Min Yuen (People's Movement). In theory, the highest authority in the Malayan Communist Party was the central committee of 11 to 15 members. In practice, however, the politbureau, through the secretary-general issued policy decisions and through the military high command gave directives or orders to the armed forces. Three regional bureaus controlled party activities in the Malay states. Below the regional bureaus were district committees. The branch, with three to five members, was the basic unit.

Typically the Communist technique of interlocking the underground with the military organization was used. A committee from the politbureau was established within the high command of the Malayan Races Liberation Army. Regional committees controlled the MRLA units in the regions. Each committee usually consisted of three members: a party representative, a military commander, and a military vicecommander (for troop indoctrination). The party representative was the controlling figure.

The Min Yuen, at first a front organization established to replace the party-controlled trade unions that disappeared when the Emergency began in 1948, became an underground organization. Its functions were to collect funds for the MRLA, furnish supplies, collect intelligence, disseminate propaganda, and serve as a pool for recruits for the party and the MRLA. In many areas the Min Yuen had its own armed forces, which protected the political organization and carried out independent terrorist activities in support of the MRLA.[48]

In the Philippines the Hukbalahap (Huk) organization was patterned after other Communist insurgent organizations. The chain of command included the secretary-general, a 31-man central committee, and an 11-man politbureau. The secretary-general directed national committees for organization, education, finance, intelligence, and the military. The military committee, made up of members of the politbureau, exercised command control of the army. Its policies and decisions were carried out by the commander of the army who, with his deputy commanders and staff, formed the GHQ.

The party's organization bureau was at the same level as the army GHQ, and was charged with handling political affairs and furnishing political and propaganda guidance to the commander of the army. Directly under GHQ were the party regional committees (RECOs), which at one time numbered about 10 and were believed to be the highest headquarters in direct command of troop units for tactical operations.

The RECOs functioned also as territorial and administrative head-quarters, with responsibility for organization and propaganda functions. Regional commanders represented GHQ in their area, supervising and coordinating military plans and political activities. They also developed tactical plans for implementation by their subordinate field commanders. At the regional level there were also organization committees in charge of establishing local Communist cells, observing the loyalty of party members, indoctrinating new members, and supervising their training.

Young men who were Huk members were sent back to their native villages to recruit for the guerrilla force. They were to organize their villages for the Huks by establishing committees of trusted and respected elders. Each village committee was organized into two groups: one a non-militant underground group, the other military. Most of the male villagers were recruited as guerrilla fighters and organized into local reserves. They worked their farms by day and served as guerrillas at night. The local underground provided intelligence, food supplies, and medical care for the guerrilla units.[49]

The Greek resistance movement during World War II provides another example of how a small but relatively well-organized party can organize a vast military and civil insurgent movement and maintain control over a large segment of the population. In September 1941, the Greek Communist Party (KKE) began a nationwide resistance movement. The Communists organized the Greek National Liberation Front (EAM) to recruit and enlist all sectors of Greek society. All citizens and classes that favored resistance against the Nazis and free elections after the war were welcomed. The EAM sought and won the support of some left-wing political leaders and formed a coalition made up of the KKE, the Union of Popular Democracy (ELD), the Socialist Party of Greece (SKE), the United Socialist Party of Greece (ESKE), and the Greek Agrarian Party (AKE). Each of these parties had one representative on a central committee. This committee was the policy-making organ of the EAM. Although it appeared that the EAM was composed of many parties, in fact the Communists held most of the power through the front's functional groups. The EAM organized the Greek People's Liberation Army (ELAS), the United All Greece Youth Organization (EPON), the National Mutual Aid (EA), and the Worker's National Liberation Front (EEAM).

The British estimated that the ELAS was composed of approximately 50,000 men while the EAM had from 300,000 to 750,000 members. The EAM gave estimates of 85,000 for the ELAS and approximately 2 million for the EAM. By 1943, the EAM controlled a good portion of Greece through an administrative hierarchy and local self-government. In each village there were four EAM groups: the EA, the EPON, the Guerrilla Commissariat or ETA, and the general EAM committee. The secretary of the EAM was called the *ipefthinos* or "responsible one." It was his duty to check on all travelers coming into a village to confirm their identification, to recommend individuals for training to the ELAS, and to follow orders from his district superior. The ETA collected taxes and gathered materials for the ELAS.

It was through the *ipefthinos* in each village that the Communists exerted control. Before the war the Communists were the only party with widespread underground experience and most of the organizers of the EAM were Communists. Thus, the Communists who organized the EAM in a village usually became the leading EAM functionary. Through the election of higher level officials, the Communists further increased their influence. Village *ipefthinos* elected the members of the district EAM committee. The district members in turn elected a regional committee. Each region in turn had one representative on the EAM central committee. Cities such as Athens, Piraeus, and Salonika also had one representative on the central committee. The city representatives were elected by neighborhood EAM committees, and EAM professional and trades organizations.

The 25 delegates on the central EAM committee represented political parties, the functional organization, the ELAS, EPON, EA, and regions and large cities. Through the *ipefthinos* the Communists had a majority of the delegates on the EAM central committee although they made up approximately one-tenth of the EAM membership. The EAM central committee appointed the two top leaders of the ELAS. The ELAS had two bodies, a high command which carried out the military operations and a central committee which passed on policy. However, all political issues were resolved or handled by the EAM. The EAM controlled the armed services through the *kapetanios*. While each ELAS unit was led by a military commander who was responsible for all military decisions, at a comparable level there was a *kapetanios* who was responsible for propaganda and morale within the unit and the relations between the units and civil population.

In the ELAS headquarters there was a military commander, a *kapetanios*, and an additional EAM representative. Invariably the *kapetanios* and the EAM representative were Communists. Although the military was in command of all military operations, the army was controlled by the EAM and, ultimately, the Communist Party.[50]

The most recent example of Communist insurgent organization and one which requires detailed description, is in South Vietnam. Again, the liberation front and the establishment of separate but interlocking organizations have been employed by the Communists. The North Vietnamese

Lao Dong (Communist) Party provided the impetus for establishing the National Front for the Liberation of South Vietnam (NFLSV), and the National Liberation Army. Collectively, these organizations are known as the Viet Cong. To promote the idea that the South Vietnamese National Front and the Lao Dong were separate organizations, and to counter the charge that North Vietnam had invaded South Vietnam, the People's Revolutionary Party (PRP) was organized. The leadership of the NFLSV, the National Liberation Army, and the PRP is an interlocking one, the Communists holding multiple positions in all three organizations.[51]

The Liberation Army receives direction from the Lao Dong, the Ministry of Defense, and the high command of the North Vietnamese People's Army. The highest military headquarters of the Liberation Army in South Vietnam is the Central Office for South Vietnam (COSVN), which is below the high command of the People's Army. The Central Office maintains lines of communications with Hanoi and the major military units in the South.[52]

The National Liberation Army is composed of regular military and paramilitary units. Within the regular army are the main forces that operate across provincial borders and the regional forces that operate within single provinces. The paramilitary force, the Guerrilla Popular Army, has three elements: the village or hamlet guerrillas, the combat guerrillas, and the secret guerrillas. These are organized in squads or platoons and are essentially part-time guerrillas who carry out harassing and sabotage missions and serve as a replacement pool for the regional units.[53]

Political direction and control of the National Front is exercised by the Lao Dong through the COSVN.[54] The NFLSV is composed of a number of organizations. Besides the Communist inner core—the People's Revolutionary Party—two other political parties are represented in the movement: the Radical Socialist Party and the Democratic Party. Then there are various liberation associations (farmers, women, youth, students, cultural, and workers) and many professional and special interest groups. Finally, there are several minor groups, such as the Afro-Asian Solidarity League and the Peace Preservation Committee, which are oriented toward external or international matters.[55]

To insure control by the party, the policy is to have at least two party members working openly on most committees. In practice, the number is usually determined by the control the Viet Cong exercises in the area. To achieve a broad base of representation in the NFLSV, it has been standing policy that party members on NFLSV committees should never exceed two-fifths of the committee's membership.[56]

Administratively, the National Front for the Liberation of South Vietnam is organized along geographic lines, with an interzone, zone, province, district, village, and cell structure. Operating at the interzone level are a committee and a committee chief. The committee is organized into six branches, with the *dan van* branch responsible for civil and military recruiting and proselytizing. The organizational branch is responsible for the *dich van* movement or armed propaganda and terrorism among the government civil service workers and the Army of the Republic of Vietnam (ARVN). This branch is charged with influencing government employees through demonstrations and riots. The *binh van* is responsible for infiltrating, building cells, and subverting the morale of government troops and encouraging them to defect. The training and indoctrination branch is charged with training *agitprop* and armed propaganda teams. A liaison and communications branch conducts the organizational communications needs through couriers and runners. The security branch is responsible for the internal security of the organization. This administrative structure is represented at each level of the organization.

The NFLSV focuses its attention on several major activities. The political struggle (seeking out and fomenting class conflict) is one of its principal activities. Another major activity is the organization and expansion of the united front groups. Finally, the NFLSV is directed to undertake "armed struggle," including sabotage, assassination, and other forms of violence and coercive persuasion.

The front is primarily charged with the administration of liberated areas, liberation associations, and the various professional and special interest groups. It also carries out missions of espionage and sabotage against the government and is responsible for action among the masses. Mass communication, ranging from radio to agitation, falls within its responsibility.[57]

At the provincial level, the front has established shadow governments headed by a commissioner who presides over a pyramidal structure topped by a central committee for the province, which supervises district commissioners, village or township cadres, and hamlet committees. Political administrators, many of whom have received from 6 months to 2 years of training in civil administration in Hanoi, govern these areas. Although many administrators are South Vietnamese, there are some North Vietnamese advisers as well.

The basic administrative bodies in the "liberated" areas are the liberation committees. There are subcommittees for military matters, health, information, culture, education, and communications. There is a Foreign Relations Commission which has missions in Moscow, Peking, Prague, Algiers, Djakarta, and Havana. In addition to the committees on the national level, there are liberation committees at the province, district, and village levels.

When a village is liberated, elections are held to choose a local committee for self-administration, which then establishes groups for military matters, public health, education, and economic questions. The village committee elects representatives to the district committees, who in turn elect delegates to the provincial committees. However, these higher committees were initially appointed. Elections are to be held at some later date.

Although village committees can contact district and provincial committees for advice, the higher committees usually provide the general political line and leave local affairs to the self-administration committees. One reason for this is the difficulty of maintaining daily communication among the various administrative units. Messages usually come from broadcasts of the clandestine Liberation Radio or circular letters.

For training purposes, books are distributed by the provincial education committees. These committees have crude jungle print shops in which most of the work is done manually. In addition to standard textbooks, books are also printed for the national minorities. Each village sends people to the teacher-training programs in the district or provincial headquarters; they then return to their villages and organize schools.

Each village also sends two people to the district center to receive a 6-month training course in medicine and hygiene. On returning to the village, they set up clinics, where most attention is paid to problems of hygiene and the control of malaria and intestinal diseases. Serious medical cases are sent to district or provincial clinics.

Because of transportation difficulties each liberated village is responsible for growing its own food, with surplus food going to the National Liberation Army.[58]

COMMUNICATIONS

In Communist insurgencies, there are several kinds of communication networks. There is a communication network within the Communist Party, another within the National Liberation Front (NFLSV), and a third within the National Liberation Army. The systems vary from the simple to the complex in operation.[59]

Within the party, disruption in communications between members of the cadre would affect operations very little, for the cadre is so well trained and indoctrinated that its members need little guidance or instruction in day-to-day operations. Only major policy changes or new programs make it necessary to communicate within the organization.

There are two types of formal communications. One is the written directive, which is relatively slow. The fast type is the radio message, communicated through clandestine stations. When radio is used, the message is disguised through some type of code, often in the form of key words or phrases. Intensive training and preplanning reduce the need for fast communications.[60]

If a message is of extreme complexity or great secrecy, a meeting may be arranged. The lower echelons receive word by courier that a meeting is to be held. A liaison member from higher headquarters outlines the wishes of the central committee and entertains questions at this meeting. He may cite and quote documents which he has read and burned. To maintain maximum security, secret documents are seldom carried to meetings and no notes are taken.

Administratively, there are four levels at which such meetings take place —the national headquarters of the central committee, the provincial central committee headquarters, the district or town central committee, and the cells. Each central committee has as part of its administrative operation a communications liaison section. Each of the lower echelons has its own liaison with its parallel units. The communications section handles all messages, coding and decoding them, and supervises the couriers.

The communications system of the NFLSV operates much as the party system does, but with less emphasis on security and a greater volume of messages. There are message liaison centers at central committee headquarters, interzone headquarters, zone headquarters, province headquarters, district headquarters, and village headquarters. Messages may concern new missions and assignments, reviews of policies and programs, or distribution of publications and propaganda.

In discussing communications within the National Liberation Army, it is important to distinguish between the regular mobile main force and the paramilitary units. The guerrilla units operate out of the villages and use runners to the district central committee. Messengers carry communications between the lower echelon paramilitary units and their political counterparts without going through headquarters. The guerrillas communicate with the regular forces by runners, although they keep contacts to a minimum. Most communications between these two originate within the NFLSV.

Main-force communications are more sophisticated, depending upon the type of equipment captured from government forces. At headquarters, captured radios and walkie-talkies may be used, as well as couriers and messengers. Since the main body of central headquarters is dispersed, telephone systems and switchboards are used to coordinate central committee operations. Communications systems are extensively used as an alert system. Field telephone, telegraphs, and lights or visual signals are used to warn the main force of enemy approach. During operations, communications are generally limited. In the "liberated" areas a regular telephone and telegraph system may be used.

A National Courier Division is charged with running message between the regional central committees and the district committee or cell level. Courier services are usually a system of liaison and jungle mail-drops. Training for couriers, who are frequently "innocents" (children, girls, women), is detailed. A courier for district headquarters spends approximately a year within the organization before being given his assignment. The district headquarters controls as many as six branch committees and platoons which work on the edges of the jungle or in rural areas. The courier's job is to spend his days visiting "jungle letterboxes" collecting and delivering messages. Letterboxes are the only fixed point of contact between regional headquarters and the district branches and platoons. Couriers seldom meet each other, leaving messages on different days in order to maintain security. Letterboxes are also changed from time to time and their location is secret.[61]

While the party is highly secretive, limiting communications to only the most vital, the National Liberation Front (NFLSV) has more extensive communications, often concealing the substance in propaganda and morale messages. The Liberation Army's communications are complex on the tactical level, using modern technical equipment in addition to runners, couriers, and jungle mail-drops. Within the party and NFLSV, the communications structure consists of a series of liaison couriers and runners.

SHADOW GOVERNMENT AND POPULATION CONTROL

Communist underground strategy is to build as well as to disrupt. Through shadow governments, the Communists develop new political institutions and new symbols of authority which serve as instruments for population control. The Communists gain control of the civilian populace by combining the positive incentives of political doctrine and institutional order with the negative sanctions of terrorism and coercion. They develop institutions, such as rural courts, youth leagues, schools, and farmer co-operatives to exert normative and regulatory control over individuals, and reinforce this control with coercive means, such as surveillance, threats, and physical punishments.[62]

Shadow governments are usually initiated in towns and villages, where little or no governmental control exists. Such places frequently have none of the advantages of community action or organization, such as schools, sanitation facilities, medical services, police protection, courts, or political participation. The shadow government parallels local government structures. Through the systematic removal or assassination of government officials, and through agitation and propaganda, official government control is eroded and replaced by new institutions.

In the opening phases of subversion, infiltrators live among townsmen, carefully select young, potential leaders, and organize them into nucleus cells. Cell members then agitate and turn specific grievances into crystallized attitudes. Later, armed guerrillas and professional agitators visit the villagers. With the implicit threat provided by the guerrillas, "free" elections are held. The slate of candidates includes members from the covert cells. A shadow government is organized behind a facade of representative government. Through social and professional organizations, social control is established and initial resistance is turned into varying degrees of acceptance as villagers welcome the advent of a school, social organizations, sports clubs, and so forth. The old power structure is assaulted as drastic changes are made in the village's political and economic institutions. As a multiplicity of new organizations requires new leaders, young members of the village, usually those reunited by the underground, are given the responsibility and power usually reserved for the elders or the wealthy.[63]

As a village gives more and more support to the underground shadow government, the legal government may be prompted to retaliate or launch a counterattack. Villagers who were involved in the shadow government or who resist the government's return may be punished. However, the government's return is often short-lived, with government troops generally leaving after a brief period, because of the remoteness of seeming unimportance of the village in overall government programs. The Communists then return to resume their shadow-government control. Local commitment tends to be made on the practical basis of who is in control rather than who is preferred.

When a village has acquiesced, the underground progressively transforms it into a base. The villagers are induced, through normative or coercive means, to provide food, money, and a place to store arms and ammunition. Village men are recruited into active service, farming their land by day and conducting ambushes or raids at night. The villagers are also a valuable source of intelligence. They infiltrate government headquarters and installations and provide advance warnings, helping the militarily weaker guerrilla forces to plan offensives and ambushes or to avoid encirclement.

Population control is maintained by organizing multiple, interlocking memberships among the inhabitants. Constant social interaction and propaganda-oriented discussions are enough to convince many to support the insurgent government. Instruments of social force, such as courts and law enforcement agencies, are usually sufficient to coerce any doubters. Information from self-criticism sessions, covert agents, and political police alert the insurgents to any organized opposition.[64]

The steps through which the Communists go in establishing shadow governments are illustrated in the example of a South Vietnamese village. Captured records of a Viet Cong commander revealed some of the techniques and problems the Communists faced in gaining control of a village in Kien Phong Province in 1960.[65] The village, dubbed "XB" for security reasons, had 6,000 inhabitants and was effectively transformed from a non-Communist into a hard-core Communist combat village. Interestingly, the Diem government had done a relatively good job of administering the area, and the Communists, in order to build a mass organization, finally had to use farmers' latent interest in land to focus grievances against the government. The farmers merely wanted low rent and the right to farm the land they maintained. The government soldiers in the area collected taxes and reclaimed land from delinquent tenants for the landlords, who usually lived in the cities. At first, the Communists voiced various slogans against the landlords, but achieving little success with these, they dispatched an agitprop team to organize the village. Living in fields and marshes during the day, the team slipped into the village at night to propagandize. Government mobile troops were active in the area during the first year and largely thwarted the Communists' efforts—killing three party members and arresting over 100. Indeed, only one cadre member finally remained, and he had no local support.

Eventually, through persistent efforts, several farmers were induced to join the party by "the promise" of land. Slowly, a base of seven members was established as the beginning of the local XB party. These in turn recruited others, and eventually the cadre grew to 26 members, while the XB expanded front organizations to 30 members in the Lao Dong youth movement, 274 in the Farmer's Association, 150 in the Youth Group, and 119 in the Liberation Women's Group. They were instructed to open as many private schools as possible and to form groups and associations among professionals, tradesmen, workers, and peasants.

Within two years, 2,000 villagers had become involved in the Communist front activities. Yet, despite these apparent successes, the mass of farmers remained passive to Communist appeals for action. The party adopted a new tack, beginning with a concentrated campaign to eliminate the influence of village leaders and government security agents. By applying pressure—often physical threats—to leading villagers, they were persuaded or coerced into joining the party. When government forces came to the village with projects, they were unable to find any leader who would cooperate. Even government medical supplies were turned down. Finally, the government virtually abandoned the village in terms of civic projects. The party encouraged people to take over land and began to establish public health, sanitation, and education facilities. It even assisted in the marketing of produce.

Thus the party, by persistent persuasion and coercion, grew from a small cell to the ruling authority in village XB. Throughout this process the people in the village seemed to be motivated by individual interests rather than by Communist doctrine.

At this point an attempt was made to get active support for the National Liberation Movement from the villagers. The party saw to it that Viet Cong flags were flown from village flagpoles and frequent propaganda sessions were conducted. The party began to press the villagers into military activities. In spite of initial reluctance, the village was eventually transformed into a "combat village." Villagers were induced to set up boobytraps, build barricades, and establish defensive positions against government troops. With this, the village became a target for government attack. The more the villagers were pressed by the Communists into building defenses, the more government troops came to clean up the rebel "stronghold" (even though no guerrillas were there). The Communists had maneuvered the village into "defending" itself from its own government. Inevitably, the more the government troops attacked, the more the people turned against the government and toward the Communists. The axiom that "people learn war through war" became a fateful truism: whereas the villagers were once reluctant and hesitant in their military support of the Communists, they now had a vested interest in supporting the guerrilla units and protecting the village from the government.

In another village in Vietnam, the Viet Cong entered and let it be leaked to government troops that they were going to hold the village for 3 days. Government troops attacked the village, although the Viet Cong had left in the meantime. The net result was that villagers who were once neutral were alienated from the government.[66]

Another example of the operation of shadow governments by Communist undergrounds appears in the Huk movement during World War II in the Philippines.[67] As described by Luis Taruc, the Huk leader, the Communists found that the easiest villages or barrios to organize were those which had strong prewar organizations. Once they had a "beachhead" in a village, the principal focus of their organizational effort went into establishing a Barrio United Defense Corps (BUDC).

When a BUDC unit was established, a council was formed. The size of the council depended upon the size and importance of the barrio. Elected by secret ballot, the barrio council consisted of a chairman, a vice chairman, a secretary-treasurer, a chief of police, and directors of recruitment, intelligence, transportation, communications, education, sanitation, and agriculture. Often one man held several positions. Councils were formed only in those areas where Huk squadrons could protect a barrio and defend it against the Japanese. Elsewhere, barrio relief associations and anti-robbery associations were formed.

Organizing the BUDC in a village involved several phases. The first step was to send a contact man to find out who in the barrio were loyal, who were anti-Huk or spies, and what kind of people the barrio leaders were. If barrio sentiment was favorable, a squadron would enter and call a meeting of leading individuals and talk to them. Then a meeting of the entire village would be called and the Huk program explained. Taruc reports that these barrios were easily organized.

However, in barrios which had collaborators, spies, or were predominantly against the Huks, a contact man would identify the agents and where they lived, and watch their movements. Then a Huk squadron would surround the barrio, arrest collaborators and suspected enemy agents, take them to a public place, and on the basis of the information gathered from the people of the village, make charges against them. In a public meeting people were asked if the charges were correct or if there were any additional charges. If the people repudiated the charges, the prisoners were released. If they confirmed the charges, the Huks tried to determine if the individual collaborated under force or willingly. Traitors were usually executed, but those who cooperated under duress were lectured and asked to work for the underground.

After the trial, the whole barrio was lectured and the squadron left, returning at a later date to check on village activity. If the enemy had returned to establish control and people had collaborated, the Huk unit arrested the collaborators and took them with the squadron for 2 or 3 months. During this time they lectured them. When they felt they were convinced, they returned them to the barrio.

This type of direct military intervention was not normal procedure. Usually, organization of civilian support and population control was effectively maintained through the establishment of mass organizations and new institutional forms. To develop mass organizations, the Huks initially worked with a small number of former union organizers and professional men who had been associated with popular movements. With this nucleus they set up schools to teach the techniques of mass organization. They instructed recruits in the methods of infiltration, organizing barrios, and how to penetrate and combat the Philippine constabulary. These schools eventually trained barrio council members as well. The major functions of the barrios were to help in military operations and to develop their economy so as to provide supplies for the insurgents.

Although the Huk squadrons usually camped at a distance from the barrios to avoid drawing suspicion or attention to them, occasionally they did stay in a barrio and soldiers were assigned to various houses and families. In return, the Huks helped with household chores and assisted in farming duties.

The BUDC acted as reserves for the mobile units. For each Huk in the field, there were two in the barrios, engaged in productive work and civilian pursuits. Recruits also received military training in the barrios. Usually men on active duty were rotated back to the barrios for rest periods and the barrio reserves would spend some time with the regular guerrilla units. As Taruc concludes, "In this way we are able to build an army that was very much like an iceberg in appearance, two-thirds of it beneath the sea."

The Huk avoided encounters with the enemy that might result in punitive action against the barrio. When an ambush was to be staged near a barrio, a meeting was held with the BUDC council in order to obtain its permission.

The director of intelligence of the barrio was an integral part of the overall intelligence network. Every man, woman, and child in the barrio reported information to the director who in turn transmitted this information via runner to the guerrilla agents. The most significant contribution of the barrio intelligence was their observation and investigation of strangers, which greatly limited the enemy's use of spies and agents. A director of transportation was responsible for maintaining carriage horses and carts, and for placing them at the disposal of the Huks at a moment's notice.

There was an inner communications system among the barrios, and no one could move from one barrio to another without an appointed "connection" to accompany him. There were two types of courier systems, direct and relay. Important messages were always sent directly by courier and traveled a well-defined route. The relay system was circuitous and took considerably longer. Other devices were used for alerting barrios, such as a flash of light through an open window, flags, banners, or clothes hung on a clothesline.

A director of education was responsible for setting up schools so that the children would not have to attend schools established by Japanese occupation authorities. Health and sanitation projects were also undertaken. The director of economy and agriculture had an important position. His task was to organize the village so that a maximum amount of food was available for the Huks and the villagers and kept from the enemy. Cooperatives were launched for the planting, care, and harvesting of crops. To protect food from Japanese confiscation, Huk squadrons attacked the enemy frequently during the harvesting season to divert troops from the fields. The squadrons also helped in the harvesting. The barrio organized groups to put the food into drums and bury it or pour it into hollow bamboo poles that were hidden in the rafters of houses.

Courts and a jury system were set up to administer justice. The barrio council handled both civil and criminal matters. Cases which involved informers, spies, or traitors were generally referred to the military committee in order not to involve barrio people in the executions. In order to curtail black-marketing in food, a system of licensing and patrol was established to check on the movement of all rice. No one could sell rice without a license from the insurgents and anyone caught in black-market activities was investigated and turned over to barrio courts.

In summary, the importance of shadow governments has often been overlooked in the analysis of Communist organizational strategy. The Communists seek to not merely disrupt constituted government, but to supplant it. They employ such positive forces as organizational ties and institutional norms to exert social control. But lest these "voluntary" and "normative" appeals fail, they also employ the threat of coercion, using such means as clandestine cells for surveillance and terrorism. Through shadow governments, the Communists create effective instruments of population control that not only offer an area of support for guerrilla activity but lay the foundation for the later emergence of provisional governments.

FOOTNOTES

[1] Andrew R. Molnar, et al., *Undergrounds in Insurgent, Revolutionary and Resistance Warfare* (Washington, D.C.: Special Operations Research Office, 1963), pp. 23–26. For extensive annotated bibliographic information on works published about communism, see Thomas T. Hammond (ed.), *Soviet Foreign Relations and World Communism* (Princeton, N.J.: Princeton University Press, 1965), p. 1240, and Walter Kolarz, *Books on Communism* (New York: Oxford University Press, 1964), p. 568. For more specific information on Communist organization and strategy, see U.S., Congress, House, Committee on Un-American Activities, *Facts on Communism*, Vol. I, 86th Cong., 1st Sess., 1959, especially pp. 75–120; and U.S., Congress, House, Committee on Un-American Activities, *The Communist Conspiracy: Strategy and Tactics of World Communism*, Sections A–E, 84th Cong., 2d Sess., 1956.

[2] Gunther Nollau, *International Communism and World Revolution* (New York: Praeger, 1961), pp. 162–64.

[3] *Ibid.*, p. 201ff; and Philip Selznick, *The Organizational Weapon: A Study of Bolshevik Strategy and Tactics* (Glencoe, Ill.: The Free Press, 1960), p. 50.

[4] Molnar, *Undergrounds*, pp. 146–52.

[5] U.S., Department of State, Bureau of Intelligence and Research, *World Strength of the Communist Party Organization*, 17th Annual Report, January 1965.

[6] Rollie E. Poppino, *International Communism in Latin America* (Glencoe, Ill.: The Free Press, 1964), pp. 119–20.

[7] *Ibid.*, pp. 121–22.

[8] *Ibid.*, p. 124.

[9] *Ibid.*, p. 125.

[10] J. Peters, *The Communist Party: A Manual on Organization* (New York: Workers Library Publishers, 1935), p. 38.

[11] Australia, *Report of the Royal Commission on Espionage* (August 22, 1955); and Arthur Koestler, "The Initiates," *The God That Failed*, ed. Richard Crossman (New York: Harper and Brothers, 1949), pp. 24 and 28. See also Peter Deriabin and Frank Gibney, *The Secret World* (Garden City, N.Y.: Doubleday, 1959), and E. H. Cookridge [Edward Spiro], *Soviet Spy Net* (London: Frederick Muller, Ltd., 1955).

[12] Louis F. Budenz, *The Techniques of Communism* (Chicago: Henry Regnery Co., 1954), pp. 159–61.

[13] Selznick, *Organizational Weapon*, pp. 18–19.

[14] *Fundamentals of Marxism-Leninism: A Manual*, 2nd ed. (Moscow: Foreign Languages Publishing House, 1963), pp. 347–48.

[15] Selznick, *Organizational Weapon*, pp. 20 and 60–61.

[16] Peters, *Communist Party*, pp. 23–26.

[17] *Ibid.*, p. 24.

[18] T. Angress Werner, *Stillborn Revolution: The Communist Bid for Power in Germany, 1921–1923* (Princeton, N.J.: Princeton University Press, 1963), p. 470.

[19] Peters, *Communist Party*, pp. 32–33; and Selznick, *Organizational Weapon*, p. 33.

[20] Budenz, *Techniques of Communism*, pp. 106, 108.

[21] *Ibid.*, p. 109.

[22] Poppino, *International Communism*, pp. 117–18; see also Peters, *Communist Party*, cited in U.S., Congress, House, Committee on Un-American Activities, *The Communist Conspiracy*, Part I: *Communism Outside the United States*, 84th Cong., 2d Sess., 1956, H.R. No. 2244, pp. 120–34.

[23] Selznick, *Organizational Weapon*, pp. 31–32.

[24] Budenz, *Techniques of Communism*, p. 35.

[25] U.S., Congress, House, Special Committee on Un-American Activities, *Hearings*, Appendix, Part I, 76th Cong., 1st Sess., 1940, p. 484; cited also in Selznick, *Organizational Weapon*, p. 118.

[26] Selznick, *Organizational Weapon*, pp. 207 and 213–14.

[27] *Ibid.*, pp. 78–79.

[28] Lucian W. Pye, *Guerrilla Communism in Malaya* (Princeton, N.J.: Princeton University Press, 1956), p. 60.

[29] *Ibid.*, pp. 61 and 74–75; see also Molnar, *Undergrounds* pp. 253 and 256.

[30] E. R. Wainhouse, "Guerrilla War in Greece 1946–1949: A Case Study," *Military Review*, XXXVII (June 1957), p. 19.

[31] Molnar, *Undergrounds*, pp. 315–18.

[32] Selznick, *Organizational Weapon*, p. 96.

[33] *Ibid.*, p. 192.

[34] Budenz, *Techniques of Communism*, pp. 157–58.

[35] Molnar, *Undergrounds*, pp. 90–92; Selznick, *Organizational Weapon*, p. 106.

[36] Selznick, *Organizational Weapon*, pp. 115, 121.

[37] Budenz, *Techniques of Communism*, pp. 157–58.

[38] Pye, *Guerrilla Communism*, pp. 77–78.

[39] Selznick, *Organizational Weapon*, pp. 144–45.

[40] *Ibid.*, pp. 126–44.

[41] V. I. Lenin, "What Is To Be Done?" *Collected Works* (New York: International Publishers, 1943), pp. 422ff; see also Bertram D. Wolfe, *Three Who Made a Revolution* (New York: Dial Press, 1948), p. 122, and Selznick, *Organizational Weapon*, p. 140.

[42] Special Operations Research Office, Foreign Areas Studies Division, *Area Handbook for Venezuela* (Washington D.C.: Government Printing Office, 1964), pp. 526–28.

[43] P. J. Honey, "North Vietnam's Workers' Party and South Vietnam's People's Revolutionary Party," *Pacific Affairs*, XXXV (Winter 1962–63), p. 376.

[44] *Ibid.*, pp. 375–83; see also Hoang Van Chi, *From Colonialism to Communism* (London: Pall Mall Press, 1964), p. 51, and U.S., Department of State, *A Threat to the Peace: North Vietnam's Effort To Conquer South Vietnam* (Washington, D.C.: Department of State, 1961), pp. 10–11.

[45] Vo Nguyen Giap, *People's War, People's Army* (Washington, D.C.: Government Printing Office, 1962), pp. 98ff.

[46] Mao Tse-tung, *Selected Military Writings* (Peking: Foreign Languages Press, 1963), p. 134; for another translation see Mao Tse-tung, *Selected Works*, Vol. I (London: Lawrence and Wichart, 1954), p. 24.

[47] U.S., Congress, Senate, Committee on the Judiciary, Subcommittee to Investigate the Administration of the Internal Security Act, *Hearings on the Communist Threat to the United States Through the Caribbean*, 86th Cong., 1st Sess., 1959, p. 75.

[48] For a more thorough discussion see Fred H. Barton, *Salient Operational Aspects of Paramilitary Warfare in Three Asian Areas*, ORO–T–228 (Chevy Chase, Md.: Opera-

tions Research Office, 1953), pp. 31ff; Pye, *Guerrilla Communism*, pp. 41ff; and Molnar, *Undergrounds*, pp. 245ff.

[49] For a more thorough discussion see Barton, *Paramilitary Warfare*, pp. 50ff; Molnar, *Undergrounds*, pp. 313ff; and Alvin H. Scaff, *The Philippine Answer to Communism* (Stanford, Calif.: Stanford University Press, 1955), especially pp. 141ff.

[50] For a more through discussion see D. M. Condit, *Case Study in Guerrilla War: Greece During World War II* (Washington, D.C.: Special Operations Research Office, 1961); L. S. Stavrianos, "The Greek National Liberation Front: A Study in Resistance, Organization and Administration," *Journal of Modern History*, XXIV (March 1952), pp. 42–55; and Hugh Seton-Watson, *The East European Revolution* (New York: Praeger, 1951).

[51] Douglas Pike, "The Communication Process of the Communist Apparatus in South Vietnam" (unpublished manuscript, Center for International Studies, Massachusetts Institute of Technology, circa 1964).

[52] U.S., Department of State, *Aggression from the North: The Record of North Vietnam's Campaign to Conquer South Vietnam* (Washington, D.C.: Department of State, 1965), p. 23.

[53] Bernard Fall, *The Two Viet-Nams: A Political and Military Analysis* (New York: Praeger, 1963), p. 351; and Pike, "Communication Process."

[54] Department of State, *Aggression from the North*, p. 22.

[55] Pike, "Communication Process," p. 28.

[56] Department of State, *A Threat to the Peace*, p. 47.

[57] Pike, "Communication Process," pp. 16ff.

[58] Wilfred Burchett, *The People's Regime in South Vietnam* (Sofia, Bulgaria: Sofia Domestic Service, July 1, 1965).

[59] For additional information on the communications networks of the Communists in Vietnam, see Pike, "Communications Process," pp. 1–15.

[60] Peter Grose, "Vietcong's 'Shadow Government' in the South," *The New York Times Magazine*, January 24, 1965.

[61] William J. Pomeroy, *The Forest: A Personal Record of the Huk Guerrilla Struggles in the Philippines* (New York: International Publishers, 1963), p. 146; Luis Taruc, *Born of the People* (New York: International Publishers, 1953), p. 134; Carlos Romulo, *The Magsaysay Story* (New York: John Day, 1956), pp. 114–17; United States Operations Mission, *National Identity Card Program—Vietnam* (Saigon: Public Safety Division, U.S. Operations Mission, 1963), p. 4; Richard L. Clutterbuck, "Jungle Courier," *Marine Corps Gazette*, XLVIII (June 1964), pp. 32–36; and Richard L. Clutterbuck, "The SEP— Guerrilla Intelligence Source," *Military Review*, XLII (October 1962), pp. 13–21.

[62] Franklin A. Lindsay, "Unconventional Warfare," *Foreign Affairs*, XL (January 1962), pp. 264–74.

[63] Denis Warner, *The Last Confucian* (New York: Macmillan, 1963), pp. 104–105 and 110.

[64] *Ibid.*, p. 14.

[65] *Ibid.*, pp. 121–29.

[66] Edward G. Lansdale, "Vietnam: Do We Understand Revolution?," *Foreign Affairs*, XLIII, No. 1 (October 1964), p. 85.

[67] Taruc, *Born of the People*, pp. 117–25; see also Floyd L. Singer, *Control of the Population in China and Vietnam: The Pao Chia System Past and Present* (China Lake, Calif.: U.S. Naval Ordnance Test Station, November 1964), pp. 46–47, and Labignette Ximenés, *et al.* (eds.), *Revue Militaire d'Information*, France (February–March 1957), pp. 25–31.

PART II

MOTIVATION AND BEHAVIOR

INTRODUCTION

Historically, people have endured hardship and tyranny passively for generations. Then suddenly one generation, one segment of society, or even one individual rebels, while others continue to tolerate their traditional lot. The question of what causes such rebellions is one of the most intriguing of our times.

There are other associated questions: Are there particular factors associated with the outbreak of insurgency? Are these factors related to the success or failure of the insurgency? Who are the insurgents; are they the outcasts and malcontents who represent the criminal element of society, or are they a typical cross section of society? What circumstances and reasons motivate them to become part of an illegal, subversive movement? Once a part of the insurgent movement, what makes them persist, under hardships and danger, for the long years of struggle necessary to win a protracted war? Who are those who falter and defect?

Ideology is an integral part of most insurgent movements. Especially in the early phases when the entire organization is underground, ideology rallies people to the movement. Coupled with other organizational processes, it steels and disciplines underground members. What are the ideological and behavioral techniques of social control? How are they used in underground organizations, and what special ones are used in Communist undergrounds?

Especially important in the underground arm of the organization is the role of clandestine and covert behavior; indeed, special attention and training are usually given to teaching underground members such forms of behavior. Is there a common pattern of subversive behavior which shields undergrounders from detection and capture? What human factors techniques are used to deceive security forces?

These are some of the questions to which the following chapters are addressed. Chapter 3 presents a summary of findings about factors related to the underlying causes of insurgency, the characteristics of insurgents, and the reasons for joining, staying in, and defecting from underground and other insurgent organizations. Chapter 4 presents a summary of findings based upon theoretical studies of group behavior, relating them to qualitative descriptions of underground behavior. In Chapter 5 the rules of clandestine and covert behavior are described and related to human factors considerations.

CHAPTER 3

MOTIVATION

FACTORS RELATED TO INSURGENCY

It has been assumed that there are certain underlying causes for insurgency and that certain economic-social conditions are more conducive to the outbreak of insurgency than others.

It is commonly believed that countries rich in economic resources are not likely to be threatened by insurgency, since the basic needs of most of their citizens are satisfied. The Communists have concentrated their subversive activities on the developing nations on the assumption that a low level of economic development offers the necessary objective conditions for the creation of an insurgency and the psychological fuel to carry it through to a successful conclusion. In the 1960 conference of the Communist Workers' Parties in Moscow, Nikita S. Khrushchev advocated a "three-continent theory" which would commit world communism to the support of wars of national liberation in Africa, Asia, and Latin America. Although the Chinese Communists currently are in disagreement with Soviet policy. Lin Piao's 1965 declaration supports the contention that world revolution can be obtained through the less-developed areas of the world.

Casual observation would confirm the Communist assumption. Since 1946 insurgency has occurred in countries at all levels of economic development except the highest or mass-consumption level, as in Australia, Canada, the United Kingdom, the United States, and most Western European countries. Interestingly, however, in Latin America insurgency has occurred in those countries with high levels of development for the area—Cuba, Venezuela, and Colombia. (Haiti is an exception.)

There are few comprehensive studies on the relationship between economic factors and insurgencies. One study of world economic conditions and violence found a curvilinear relationship between gross national product per capita and political domestic violence. There was a low level of violence in countries with a high GNP per capita (more than $800) and a relatively low level of violence in those with extremely low GNP per capita (below $100), in contrast with a high rate of domestic political violence in the middle-income countries.[1] This would suggest that there is no simple relationship between economic factors and the outbreak of violence.

In order to further investigate this assumption, information was gathered on 24 recent (since 1946) insurgencies, and comparisons were made of such factors as gross national product per capita and annual increase in GNP per capita, among countries which had insurrections and those which did not. Another comparison was made to determine the relationship of these factors to the success or failure of the insurgency. (See Appendix B.)

One hundred and twenty-two countries of the world were divided into four groups on the basis of GNP per capita and a comparison made to determine how the 24 insurgencies were distributed with respect to GNP per capita. No relationship between GNP per capita and the outbreak of insurgency appeared. Insurgency occurred at all levels of GNP per capita.

People may be less prone to rebellion if the economy is improving, thus providing them with a rising level of expectations. They may feel that although they do not possess wealth now they may be able to attain it in the future. However, a comparison of the level of growth in gross national product per capita showed that insurgencies occurred in countries with high growth rates as well as low growth rates; there was no relationship between growth in gross national product per capita, and the outbreak of insurgency.

In addition, neither the GNP per capita, nor the annual increase in GNP per capita was related to the success or failure of the insurgencies in the countries studied. One explanation for the absence of relation between economic factors and the outbreak of insurgency may be that in addition to poor economic conditions, a precipitant event is required to crystallize dissent and to trigger the insurgency.

Another explanation may be that while economic factors may be important when considered on a local or regional basis or for various sub-groups within a nation, the gross national product is probably too broad an indicator to be generally valid. However on the basis of the available information and the comparisons made, the economic theory of insurgency is not substantiated by the data.

Within recent years, the occurrence of insurgency in rural areas has drawn attention to low population density as a necessary condition for insurgency. In comparing the population characteristics of nations involved in recent insurgencies, it was found that although half of these insurgencies occurred in rural, low-density population countries, a relatively high number of insurgencies also occurred in countries with high concentrations of people in urban areas. This suggests that insurgencies are not restricted to countries with rural, low-density populations, but occur also in countries with high-density urban populations. Further, there was no relation between urban-rural characteristics of the country and whether the government or the insurgents won.

There was no relationship between a country's percentage of adult literacy and the occurrence of insurgency, nor was this factor related to the outcome of the insurgency.

Since student populations seem to be involved in insurgent movements, an analysis of the percentage of students enrolled in higher education was made and it showed no relationship between the percentage of students in higher education and the occurrence or outcome of insurgency within a country. It may be hypothesized that the educational level of those within the insurgent movement itself may be important to its success or failure, but there seems to be no relationship between national literacy and higher education and either the occurrence or the outcome of insurgency.

In those countries with a high percentage of military personnel in the total population, there were more insurgencies than in other countries. It is difficult, however, to draw conclusions from the gross figures available; the percentage of military personnel per capita may have increased because of the insurgency, or the insurgency may have occurred because of the high number of military personnel.

In another study of the relationship of the size of the military establishment and deaths due to domestic violence in 33 countries, it was found that as the size of the military establishment increases there is a decrease in domestic violence. The author concludes that the motive for the creation of many large military establishments may well be the suppression of domestic dissent. It was also found that executive stability was negatively related to the size of armies in these countries and that there was a more rapid rate of turnover in countries with large armies than with small ones.[2]

1. *Insurgency and the stage of economic development.* A country's stage of economic development provides no immunity to insurgency. With the exception of the few mass-consumption societies, insurgency has occurred in countries at all levels of economic development. While economic factors may be important when considered on a local or regional basis or for subgroups within a nation, neither gross national product per capita nor GNP per capita increase is related to the outbreak or the success of the insurgency. There may, however, be some relationship between GNP per capita and the level of violence.

2. *Rural versus urban insurgency.* Insurgency is not restricted to countries with rural low-density population; it also occurs among urban, high-density population countries. Further, no relationship is apparent between the rural-urban pattern and the success of the insurgency.

3. *Literacy and education.* Insurgency occurs in countries with both high and low adult literacy. While it may be hypothesized that the educational level of those within the insurgent movement affects its success or failure, there seems to be no relationship between national literacy or higher education and either the occurrence or the outcome of insurgency.

4. *Percentage of military personnel.* No clear evidence appears as to the relationship between the percentage of military personnel in the population and the occurrence of or the success or failure of an insurgency. However, there is a decrease in domestic violence in those countries with large military establishments and no insurgency. There is also less executive stability and a rapid turnover in countries with large military establishments.

WHO ARE THE INSURGENTS?

How widespread is the insurgency in terms of the number of people actively involved? In a study of undergrounds in seven countries, it was found that only a small percentage of the total population actually participated in the movement. (See table I.) Considering the peak percentage of combatants—both insurgents and government security forces—it was found that from 0.7 to 11 percent, with an average of 6 percent, of the population were directly or indirectly involved. However, there may have been sympathizers within the rest of the population.

The ratio of underground to guerrilla personnel ranged from 2 to 1 to 27 to 1, with an average of 9, indicating that a large proportion of insurgents work at everyday jobs and only 1 out of 9 joins a guerrilla group.

Table I. Ratios of Underground Members to Guerrillas and Combatants to Population for Seven Insurgent Movements [3]

Country	Underground guerrilla	Percent of combatants* in total population
France (1940–45)	3/1	2.2%
Yugoslavia (1941–45)	3/1	2.9%
Algeria (1954–62)	3/1	8.3%
Malaya (1948–60)	18/1	8.1%
Greece (1945–49)	27/1	11.2%
Philippines (1946–54)	8/1	0.7%
Palestine (1945–48)	2/1	7.0%

* Underground, guerrilla, and security forces.

In summary, since so few of the total population participate, insurgency can be described as a low-intensity conflict in which the active combatants make up only a small proportion of the country's population. Most of the combatants maintain themselves by performing their "normal" functions within the society along with their clandestine underground activities.

Documented evidence on the place of origin of the insurgents is scarce. Available data are based on interrogation of captured insurgents, on interviews with local civilians, and on security forces' intelligence records.

In the Malayan (1948–1960) and Philippine (1946–1954) insurgencies and the Korean conflict,[a] 70 to 80 percent of the members of the movements were native to the provinces in which they operated; of these, 60 percent were native to the settlements in which they were active.[b]

Although men constitute the majority of insurgents, sizable numbers of women are active in insurgent movements.[6] In Korea, women played an unusually prominent role in the urban People's Self-Defense Units, which provided for a company of three platoons of women. About 26 percent of

[a] Although the Korean conflict (1950–1953) was a limited-war situation, there was a significant degree of guerrilla activity behind the lines. The references herein to Korea refer to the organization and operations of the Communist guerrilla forces and the supporting covert underground units.

[b] This gave the insurgents certain advantages: they were acquainted with the terrain and accustomed to its hardships; relatives and friends offered various forms of aid and assistance; their knowledge of the local people made it easy for them to pose as "neutrals" when questioned by government forces. It also gave them a feeling of fighting for their own homes and local interests.

In contrast, only 30 percent of the counterinsurgency forces came from the same province in which they operated, and of this number only 35 percent came from the same vicinity.[4] The counterinsurgents thus lacked the advantages of having people of local origin. But when they do not have old friends and relatives in the local population, soldiers are less reluctant to enforce unpopular measures.[5]

all prisoners held in Korean Government detention camps were female. Of 4,039 guerrillas captured from December 1, 1951 to January 2, 1952, about 32 percent were women. The percentage of women in the underground organization was probably even larger. In Malaya the underground arm of the insurgency, 5 to 15 percent of the Min Yuen, were women who acted as couriers, agents, and occasionally as saboteurs. In the Philippine Huk uprising, a few women held high positions in the movement's leadership.

In Korea, Malaya, and the Philippines, most age groups were represented. The underground noncombatant units had proportionately more old people and young people than did the guerrillas. (See table II.)

Table II. Percentage of Guerrilla and Underground Members
in Three Insurgencies: Korea, Malaya, Philippines [7]

Age	Guerrillas (%)	Underground Members (%)
12–15	1	5
15–20	8	7
20–25	18	32
25–30	29	22
30–35	25	13
35–....	19	21

In a sample of 2,700 Korean insurgents, more than half were below age 30.[8]

Armed boys and girls 15 years old and under were captured; children 11 and 12 years old were used as regular couriers between guerrilla units and in other underground activities. It is significant to note that there is a difference in the age distribution of male and female guerrillas. While 19 percent of the male guerrillas were either below 17 or over 40, 50 percent of the female guerrillas were in this age group. (See table III.)

Table III. Age and Sex of Korean Insurgents [9]

Age groups	Male insurgents (2,000) (%)	Female insurgents (700) (%)
Below 17	5	26
18–30	49	30
31–40	32	20
Over 40	14	24
Total	100	100

According to British data on 2,000 captured Malayan insurgents, 63 percent were under 30 years of age. The data sheets on 1,300 guerrilla and underground prisoners and information on about 230 killed bandits indicate that the younger people (12 to 15 years old) and the older people (over 35) were more often active in underground operations than in the guerrilla fighting units.[10] (See table IV.)

Table IV. Age Breakdown of Captured Malayan Insurgents [11]

Age groups	Percent
15–17	9
18–25	28
26–30	26
30–35	22
Over 35	15
Total	100

Fragmentary data on 600 members of the Hukbalahap in the Philippines supplemented by information from 62 Huks resettled through the government's Economic Development Corps (EDCOR), show age distribution similar to that found in Korea and Malaya.[12]

In Vietnam, a study of the age of Viet Cong members showed that well over half (81 percent) were under 29 years of age, with the average age being 23.8 years. The study was based upon captured personnel history records of the Viet Cong's 261st Battalion. The average age of the cadre varied with rank. The assistant leaders averaged 24.6 years old, squad leaders 26.8 years, platoon-grade cadre 30.8 years, and company-grade cadre older than 30 years.[13] The greater age of the higher ranking officers probably reflects the long, intensive training which the Communist cadre receive before reaching ranking positions.

In Korea, the Philippines, and Malaya the occupational backgrounds of insurgents were similar. The bulk of the membership consisted of peasants and workers. In the town centers and urban areas, more industrial workers and members of the intelligentsia were found, and these had great influence on planning and policymaking.[14]

A survey of captured Korean insurgents described a large majority of the insurgents as coming from working-class or peasant background. (See table V.) In Malaya, it was estimated that 70 percent of the Chinese members of the guerrilla arm of the movement were from the working classes. This group formed the rank-and-file. Those classified as intelligentsia and professionals were primarily engaged in political activities and held medium or high-level executive posts.[16]

Table V. Occupational Breakdown of Korean Insurgents [15]

Occupational Group	%
Laborers and artisans	40
Farmers and peasants	30
Students	20
Former municipal employees	5
Former police employees	5
Total	100

In Indochina, additional data from a French poll among Vietminh prisoners taken during the Indochinese war shows that 46 percent were peasants and laborers, with laborers predominating; 48 percent were classified as "petty bourgeois"; and 6 percent came from the trades and miscellaneous professions.[17]

In summary, that small percentage of the total population involved in insurgency can be characterized in several ways—

1. *Type of insurgent.* The insurgent organization is composed of an underground and a guerrilla force, with an average ratio of nine undergrounders to every guerrilla.

2. *Local origin.* Most insurgents are assigned to units in areas to which they are native, while counterinsurgent forces are usually not native to the area in which they fight.

3. *Sex.* Although most of the insurgents are men, a relatively large proportion of women participate. There is a tendency for women to be members of the covert underground elements rather than to serve with guerrilla units.

4. *Age.* The membership of the insurgent movements surveyed is generally youthful; however, this tends to reflect the age pattern of the population involved. The youngest and oldest members appear most often in support activities. The average age of the cadre is generally older than that of the members of a unit.

5. *Occupation.* The occupations of the insurgents tend to be similar to those found in the area within which they operate. Units in the rural areas are composed mostly of farmers and peasants; those in the urban areas are made up of workers and intelligensia. In general, the occupation of the insurgents tended to reflect the occupational pattern found in the population as a whole.

REASONS FOR JOINING

As an insurgency escalates in size and scope, the kind of individual it attracts and the nature of motivation for joining changes. In the early stages of an underground movement, recruitment is selective. Recruits are thoroughly screened and tested for leadership potential and dedication. At the later stages of expansion and militarization, the underground usually aims for mass recruitment. (See chapter 6.) Although there is little information on why individuals join the underground during the early stages, several motivational factors can be assumed from the underground recruiting technique itself. In its initial recruiting of cadre, for example, the underground specifically looks for those with ideological sympathies. Rational ideological considerations undoubtedly influence early recruits.

Motivation for joining during the expansion and militarization phases is better documented. However, the motivation for joining an underground movement during the militarization phase is typically complex, with no single reason dominant. In addition, motivational data are usually based on interviews and records from prisoners and defectors. Prisoners being interrogated are likely to feel at the mercy of their captors and may tend to conceal their true motives and opinions and may give false answers to save face. Furthermore, the more dedicated members of the movement are less likely to defect and thus may not be represented in the available samples.

A study of the Philippine insurgency suggests that individuals usually join as the result of a combination of factors—most often reflecting immediate needs and situational constraints. A chance to obtain personal advantage—ownership of land, leadership, or position of authority—was frequently cited. Situational problems, such as family discord, violations of minor laws, and so on, also influenced decisions to join. First contact with the movement usually came through chance. An individual joined if it filled a personal need or served as an escape, or if social pressure or actual force were applied. Once he was in the movement, indoctrination and other organizational processes helped him to rationalize his commitment.[c]

This chain of interlocking acts eventually led to full-fledged membership. While not invariable, this process was typical of most of the ex-Huks interviewed.[19]

A survey of a group of captured Vietminh showed that 38 percent of the prisoners expressed a belief in the Vietminh cause.[20] Yet only 17 percent of a Huk sample of 400 prisoners expressed sympathy with the political objectives of the Communist Party.[21]

Promises and propaganda appear to have been involved in a number of cases, although their actual effects are difficult to determine. One source has noted than among ex-Huks a majority had joined the movement without any noticeable propaganda influence; most had been primarily concerned with issues like land distribution and lower interest rates. Less than 15 percent gave their only reason for joining the Huks as propaganda or verbal persuasion, although 27 percent reported persuasion or propaganda as a contributory factor. Thirty-eight percent of the Huks became involved through personal friends. It was probably later, after being exposed to propaganda and indoctrination, that motives for joining were related to specific grievances.[22]

A number of prisoners claimed that they were coerced into joining the movement, but because they were prisoners their claims may have been exaggerated. At any rate, in one study 25 percent of the Vietminh prisoners stated that they had been forced to join against their wishes and had resented being coerced. Another 23 percent also claimed to have been forced to join, but did not appear to resent the fact.[23]

---

[c] A former Huk exemplifies the interplay of motives. This man was recognized as a leader in his *barrio* and became the local contact for the Huks. He got involved in the arrangement of a rally. When the local people then decided to form a Huk unit, they chose him as its leader. He couldn't refuse without antagonizing the Huks nor could he leave the town and move elsewhere. The Huks sent him to a "Stalin University," where he was exposed to Communist thinking and propaganda, and the propaganda points which particularly impressed him were promises to eliminate usury and government corruption and to distribute land to the poor.[18]

78

Huk leaders realized that recruiting could not wait for the slow process of persuasion and free decision. In one study it was found that 20 percent of ex-Huks had been forced to join at the point of a gun or because of threats of violence against their families; [d] for another 13 percent, violence was one important factor among several others.[25]

During the Indochinese War some cases were reported of young men who had been forced to join the Vietminh by direct physical coercion. Other men entered the movement because of indirect pressure on their families or on village leaders to provide recruits. The Communists in general combined strong-arm and other pressure techniques with propaganda appeals stressing independence. They generally avoided open appeals for communism.[26]

Coercion alone did not seem to be a large factor (20 to 23 percent) in either the Huks or Vietminh. Coercion combined with other positive incentives related to personal and situational factors, however, accounted for a larger proportion of joiners (33 to 48 percent). Another important factor was the action of government troops. Of the 95 ex-Huks interviewed, 19 percent said they joined the Huks because of persecution or terrorization by government forces. The effort of the army to suppress the revolt apparently was a factor in leading many to join the movement.[27]

Reviewing the reasons stated by captured members for joining a movement, one finds a paucity and ambiguity of data and further difficulties in the interpretation of the data available. Nonetheless, certain conclusions may be stated.

1. *Multiplicity of motives.* Usually, more than one motive is present when a member joins. A combination of factors is usually cited, with no one factor being preeminent.

2. *Personal and situational factors.* Most of the motives cited for joining tend to be related to situational or personal problems and to reflect the individual's immediate needs.

3. *Belief in the cause or political reasons.* Only a minority admit that political reasons or sympathy with the Communist Party are related to joining.

4. *Propaganda and promises.* Few join because of propaganda or promises alone. These are apparently more effective when combined with situational factors.

5. *Coercion.* Coercion alone is a small but important factor in joining.

6. *Coercion with other positive incentives.* Combined with other positive incentives related to personal or situational factors, coercion accounts for a significantly large number of recruits.

7. *Government persecution.* This factor, real or imagined, appears to be a small but significant factor leading individuals to join the movement.

---

[d] One type of coercion is seen in the example of a farmer's young son who, while working in the fields, was asked by the Huks to help carry supplies to their mountain hideout. Since they were armed, he complied. After they got to the mountains, the Huks told him he had better stay or else they would report him to the constabulary, who would punish him for helping them.[21]

Although there are few empirical studies of insurgent motives for remaining with the movement, a review of two studies of use conventional military personnel may provide insights into the motives of men in combat situations and into the sustaining role of ideology.

In a study of American soldiers during World War II, it was found that the soldier's willingness to fight was not significantly affected by indoctrination, ideological justifications, or by receiving awards for exceptional valor. More important were the norms of conduct developed in small, intimate group associations with other soldiers.[28] The concern for what his fellow men within the unit thought of him was an important influence on his performance and group effectiveness. It was concluded that most non-professional soldiers fight reluctantly and are probably motivated by status-group considerations.

Another study, based on the collapse of the German Army in World War II, found that in those units which did not surrender, values such as honor and loyalty had created a sense of obligation among the soldiers. Loyalty to their comrades was more important than ideology in their willingness to continue fighting to the end. Ideology, however, did play an indirect role. The type of leadership had a positive effect upon the combat effectiveness and commitment of the individuals within a unit. When the men in the units accepted the leadership of officers and noncommissioned officers who were devoted Nazis, the units' performance was much more effective than that of units without ideologically oriented leaders.[29] If the leadership is ideologically oriented, the units seem to be more cohesive and effective, even if the members are apolitical.

In most military units, individuals fight less because they agree with the political system than because they feel a loyalty to their fellow soldiers. They develop an esprit de corps and, in spite of adversity, try not to let their comrades down. Many insurgents who have defected still have favorable memories of the comradeship and togetherness of the guerrilla camps or the underground cells.

In one study of the Philippine insurgency, it was concluded that although people joined the Huks for various reasons, there was a tendency for a person, once a member of the movement, to gradually develop new motives for staying. Members stayed on because they were made to believe that the movement would bring about a better life for them and for the masses.[30]

Insurgents often are influenced by their own propaganda and agitation themes. The impact of agitational slogans was shown in one study of 400 captured Huk guerrillas: 95 percent asserted that their main reason for fighting was to gain land for the peasants.[31]

Psychological methods and morale-sustaining techniques have been used to induce loyalty. Since defections often occurred after serious losses, the Viet Cong went to elaborate lengths to keep up morale. Those killed in battle were carried away, often by special volunteers for that purpose, and buried with great ceremony. If it was not possible to carry the dead away immediately after the battle, the insurgents returned for them at night. This experience built up support for the movement through a desire to avenge the deaths of comrades and was apparently a significant psychological factor in keeping up morale.[32]

New recruits or suspected individuals are not usually given tasks of responsibility and are kept under close surveillance. They are not allowed to leave the camp area alone. Most underground movements require recruits to take an oath promising to remain with the movement on the penalty of death. Terror or enforcing squads are also used to retaliate against defectors. Threats of revenge are especially effective when it is difficult to defect to safe areas.[e] Atrocity stories about how the government mistreats defectors are also used.

The Communists' frequent criticism and self-criticism sessions act as a form of catharsis and permit members to voice fears and problems. In this manner members may speak out and be heard. No matter how limited and directed it may be, this process apparently serves as an outlet for emotions which might otherwise lead to defection. In addition, an individual who is disillusioned with the movement will find it difficult to conceal this in the frequent self-criticism sessions.

Another significant factor which prevents people from leaving a subversive movement is the human tendency to inertia: to do what is customary and expected of them in spite of any displeasure with the organization.

Several conclusions can be drawn as to why insurgents tend to stay with the movement.

1. *Changing motives.* Motives for remaining within the movement are usually quite different from those for joining. Indoctrination and propaganda expose the individual to new ideas, of which he may have been unaware before joining. New friends and organizational responsibilities are also motives for staying.

2. *Group norms.* Insurgents are influenced by other members of the movement. They are probably more motivated by what their friends and comrades think of them than by any ideological considerations and tend to stay out of loyalty to them.

3. *Ideology.* Ideology plays an indirect role. Units whose leaders are ideologically oriented are more cohesive and effective than those whose leaders are not, even when the members of the group are apolitical.

4. *Morale-sustaining techniques.* Various psychological techniques are used to maintain morale, such as special ceremonies and group discussions that give members an opportunity to air their emotional problems and receive group support and reinforcement.

5. *Surveillance and threat of retaliation.* Continual surveillance and threats of retaliation from terror enforcement units keep many members within the movement.

---

[e] In Vietnam, insurgents controlled much of the countryside; even government-controlled hamlets were vulnerable to the Viet Cong; there were political agents in the villages; and the countryside was patrolled by small Viet Cong units who shot deserters or persons suspected of cooperating with government forces.[33]

6. *Inertia*. Simple inertia and habit may be stronger than any inclination to leave. It is easier to continue a habit than to change it.

REASONS FOR DEFECTION

Disaffection may result in a person's leaving or defecting from an insurgent movement to the government side. He seeks the easiest and safest avenue of escape. If circumstances are such that he can simply leave, he will likely do so; if, on the other hand, the possibility of going to government forces arises first, and is relatively easy and safe, he may defect. The process of leaving is, of course, an unseen phenomenon. Only defection to government forces is recordable.

Little systematic research has been done on the motivation of insurgent defectors. However, there have been case studies based on interviews with defectors from the Vietnam insurgency, the Huk rebellion in the Philippines, and the Malayan insurgency.

In Vietnam, in January 1963, President Ngo Dinh Diem began the *chieu-hoi* (open arms) program. Viet Cong defectors were offered amnesty and assistance after a short indoctrination and retraining course. Between February 18 and June 25, 1963, 6,829 Vietnamese defectors took advantage of the *chieu-hoi* program. (See table VI.)

Table VI. Viet Cong Defectors and Area of Defection [34]

Area	Number of Defectors
Central Lowlands	779
Highlands	128
Capital City (Saigon)	13
Eastern Area	1252
Western Area	4657
Total	6829

The defections from the Central Lowlands were primarily from Quang Ngai and Binh Thuan Provinces, both sparsely populated; there were none from the city of Da Nang. The majority of the defections from the Central Highlands were in the rural provinces of Darlac, Lam Dong, and Phu Bon. There were no defectors reported from the city of Da Lat and only 13 from Saigon.

The largest number of defectors, more than twice as many as from the other areas combined, came from the western region. The majority of these defections were in the An Giang Province. The wide range in the number of defectors from the various areas probably reflects local political, social, and psychological conditions, as well as Army of the Republic of Vietnam's military strength in particular areas.

Figure 8. *Areas of Viet Cong defection February–June 1963*

Central Lowlands
1 QUANG TRI
2 THUA THIEN
3 QUANG NAM
4 QUANG TIN
5 QUANG NGAI
6 BINH DINH
7 PHU YEN
8 KHANH HOA
9 NINH THUAN
10 BINH THUAN

Central Highlands
11 KONTUM
12 PLEIKU
13 PHU BON
14 DARLAC
15 TUYEN DUC
16 LAM DONG
17 QUANG DUC

Eastern Vietnam
18 PHUOC LONG
19 BINH LONG
20 TAY NINH
21 BINH DUONG
22 PHUOC THANH
23 LONG KHANH
24 BINH TUY
25 PHUOC TUY
26 BIEN HOA
27 GIA DINH

Western Vietnam
28 HAU NGHIA
29 LONG AN
30 KIEN TUONG
31 DINH TUONG
32 GO CONG
33 KIEN HOA
34 VINH BINH
35 VINH LONG
36 KIEN PHONG
37 CHAU DOC
38 AN GIANG
39 PHONG DINH
40 BA XUYEN
41 BAC LIEU
42 CHUONG THIEN
43 KIEN GIANG
44 AN XUYEN
45 KIEN GIANG (Phu Quoc)

128 — TOTAL NUMBER OF DEFECTORS

MILES
International Boundary
Province Boundaries

17th PARALLEL
Da Nang
LAOS
CAMBODIA
Da Lat
SAIGON
South China Sea
CON SON
North

779
128
1,252
4,657

**AREAS OF VIET CONG DEFECTION
in the Republic of Vietnam**

February – June 1963

83

One explanation for the low defection rate in the more populated urban areas, where it would appear to be easier to defect, could be that underground members in the cities do not have to endure the hardships that units in the field do.

An analysis was made of the relationship between RVN Government appeals and the number of Viet Cong defectors. (See table VII.) On the basis of this sample, there appears to be a low relationship between appeals for defection and the number of people who defect. Of those who defected as a result of government appeals, most heard of defection appeals indirectly from civilians and other insurgents. It is likely that many individuals decided to defect first and then became sensitive to propaganda appeals. A large number found their reasons for defection so compelling that they defected without ever having heard any appeals.

Table VII. Appeals and Defection of Viet Cong [35]

Returned of own will	210
Responded to direct appeal	48
Responded to indirect appeal	124
Total	382

In an analysis of 382 Viet Cong defectors, figures for defectors from guerrilla units were higher than those for political defectors. This is significant in that there are usually far more underground members and liaison agents than guerrillas.[36]

Table VIII. Type of Viet Cong Defector [37]

Guerrilla	189
Political	69
Liaison	19
Desertees, Draft Dodgers	91
Detained by Viet Cong	14
Total	382

There are several explanations for this disproportion: political units have a less rigorous physical existence than military ones; political units' day-to-day activities require them to reiterate propaganda themes and carry on persuasive arguments in favor of the movement, so that they tend to be insulated from thoughts of defection. The number of defections of liaison agents suggests that, in spite of insurgent efforts to place only the most reliable people in such positions, it is a highly vulnerable job. The liaison agent has unusual opportunities for defection since he usually travels by himself and goes into government-controlled areas. Amnesty offers probably influence liaison agents to take advantage of their chance to escape.

In October 1964 the *chieu-hoi* program reached its low point, with only 253 Viet Cong defecting during that month. Defection continued at this low level until April 1965, when 532 defected. Then the figures began to climb: 1,015 defectors were reported for May and 1,089 for June. The increase in the defection rate coincided with and was largely attributed to the stepped-up Viet Cong conscription program. The young men pressed into service did not have the ideological conviction of earlier recruits and, in many cases, resisted recruitment. Among the new conscripts who defected, personal hardships and the contempt shown them by the veteran Viet Cong were among the reasons cited for defecting.[38]

A 1965 study of 1,369 men and women who defected from the Viet Cong showed that most attributed their defection to the harshness of material life in the Viet Cong. Food shortages and limited medical supplies were most often mentioned. Almost none of the defectors mentioned ideological factors.[39]

In addition to material and personal factors, the military situation also affected the decision to defect. Members of guerrilla units, for example, were found to be most susceptible to defection appeals immediately after a battle—especially if their unit had suffered heavy losses.[40]

An analysis of defection from the Huk movement in the Philippines has also been made.[41] Several motives for leaving were given by the 95 former Huks interviewed, just as they gave several reasons for their earlier entrance into the movement.

Sixty-one percent gave physical hardship as their chief reason.[f] In particular, they complained about the cold, hunger, and lack of sleep. The government forces contributed to these hardships by frequent attacks. Many of the interviewees said that they were tired of years of being fugitives and just wanted to live in peace. Forty-five percent said they defected because of the failures and disappointments of the Huk organization. Specifically, they resented the strict discipline in the movement and found orders distasteful, or had lost the feeling of progress and foresaw failure of the insurgency.

Twenty-three percent surrendered because of promises and opportunities offered by the government. The most effective promise was that of free land. Mentioned almost as often was the promise that the surrendering men would not be tortured. Other promises cited were those of a job, of payment for surrendered firearms, and of freedom for those against whom no criminal charges were being held.

Almost half (45 percent) of the defectors had heard of the government-sponsored Economic Development Corps (EDCOR), and most of these indicated that the program was influential in their decision to give up. Some said the EDCOR program gave them hope for a new life.[42]

Not more than 5 percent said they surrendered because of pressure by their families. Of course, most of the Huks were unmarried young men who did not have many family responsibilities.

In sum, there was no single overall motive for the defections. The hard-

---

f The reasons for surrender given in the following discussion add up to more than 100 percent because more than one reason was given by defectors.

ships of existence and the constant pressure of pursuit, disillusionment with the Huk organization, and government promises appeared, in that order, to be the main reasons for surrender. Thus, the government effectively pressed the Huks toward surrender by maintaining steady pressure against them and by various promises and opportunities—in particular, EDCOR.

Interviews with 60 former Communist insurgents in Malaya indicated that some of the reasons that led them to join the Communist movement were related to their later defections.[43] Many joined as an avenue for personal advancement and security. They saw the party as a strong organization which would give them a voice in the future. But as they perceived the party to be growing weaker, they felt that they had made a mistake and wanted to extricate themselves as expeditiously as possible.

Most defectors gave no serious thought to leaving the movement during their first year, being too strongly involved in party work or still having high expectations. The critical phase for most came about a year and a half after joining the party. At this point they gave critical thought and reappraisal to their current position and possible future. Most had made great sacrifices for the party, and it was increasingly clear that greater sacrifices were to be demanded even while the chances of victory grew slimmer.

Many began to feel that the future was hopeless and passed through a period of doubt in which various "crises" arose that often triggered defections. One category of crisis centered on the member's inability to meet the requirements of party membership. Generally, an individual who developed personal difficulties within the party simultaneously developed critical arguments against the party's goals and methods and the Communist cause in general. Most defectors specified communism as "bad" in terms which were most meaningful in a setting of personality politics. If the party and its leaders were seen to be "corrupt," the defector could justify his own personal position and his subsequent defection.

Another category of crises resulted from the party's attitude toward sex. Party members were supposed to lead chaste lives even though 10 percent of those in the jungle were women. Even thinking about sexual matters was classified as symptomatic of "counterrevolutionary" attitudes. Permission to marry was generally refused. The party made death the penalty for rape, which was loosely defined and judged by party leaders rather than by the woman. Accordingly, the women tended to attach themselves to the party leaders, and members resented this departure from the party policy of "equality."

Another type of crisis appeared when the party failed to satisfy the defector's personal hopes, either not meeting his needs at all or doing so at too high a price. After World War II, the living standard of the general population increased and social stability improved in Malaya. In contrast, the party member often saw his life as rigged and unrewarding. Among the immediate problems cited by many ex-Communists were that they had to work too much, that life in the jungle was too boring, or that they underwent too much physical suffering. Almost one-third felt that their existence had become too dangerous. Some were pushed into a decision to defect by the death of a friend.

In nearly all cases, the decision to defect took place after several minor crises. The likelihood of a crisis leading to defection was especially strong if earlier crises were not resolved. The psychological preparation for defection was complete when a member began to formulate general criticisms of communism. Once members became disaffected, they sought to disengage themselves from the party as rapidly as possible.

Although these interviewees, because they were defectors, are not necessarily representative of the whole party membership (many of whom might conceivably fight to the bitter end rather than surrender), the Malayan data strongly indicate that there is a continuity in the defectors' motivations. At one point in time they joined the movement, at another they deserted it; the roots for both actions lay in the same purposes and hopes. As conditions changed, the attractiveness of the alternative paths for achieving these hopes also changed. As the overall prospects of the rebellion changed, many felt their desire for personal security and social advancement could be better fulfilled by defecting than by staying in the movement. Thus, apparently contradictory actions (joining and defecting) had a motivational consistency.

Certain generalizations can be made about acts of defection among insurgents in general and among members of the underground and of guerrilla forces.

1. *Types and rate of defection.* Once the individual becomes disaffected, he may stay in the movement but not participate actively, he may leave the movement simply by withdrawing, or he may defect to the government side. The rate of defection varies widely, with a high rate in some areas and a low one in others. Local factors chiefly determine the rate of defection.

2. *Multiplicity of reasons.* Defectors usually give many interrelated reasons for their defection, usually involving personal and situational factors.

3. *Conflict and crisis.* Internal conflicts and personal crises within the organization usually precede defection. Conflicts usually arise over frustration of individual goals, harsh discipline, or lack of advancement.

4. *Time of defection.* Young recruits who are forcibly conscripted tend to defect early; those who join for ideological reasons tend to reconsider and have second thoughts some months (approximately a year) after joining the movement. It is at this time that they are most susceptible to defection.

5. *Appeals.* Although many defectors are unaware of government appeals and rehabilitation programs, these programs appear to be an influencing factor among those who do hear of them.

6. *Underground defection.* There are some unique characteristics related to underground defection. There is less defection to the government

among members of guerrilla units; similarly, there is less defection in the populated urban areas than in the rural areas. There are several reasons for this: political activities probably insulate underground members from thoughts of defection; the underground is not exposed to the rugged, harsh existence of guerrilla life; and while defection may be the only option of guerrillas, underground members may be able to simply withdraw or be passive.

7. *Guerrilla defection.* Among the guerrilla units, the rigors and hardship of life in a guerrilla unit, such as bad weather and lack of food and sleep, are often cited as reasons for defection. Usually, however, a personal crisis involving individuals in the guerrilla force is the ultimate triggering force. Defection is also frequent immediately after battle, especially if there have been heavy losses among the guerrillas.

FOOTNOTES

[1] Bruce M. Russett, *et al., World Handbook of Political and Social Indicators* (New Haven: Yale University Press, 1964), pp. 306–307.

[2] *Ibid.,* p. 319.

[3] Andrew R. Molnar, *et al., Undergrounds in Insurgent, Revolutionary and Resistance Warfare* (Washington, D.C.: Special Operations Research Office, 1963), pp. 13–16.

[4] Fred H. Barton, *Salient Operational Aspects of Paramilitary Warfare in Three Asian Areas,* ORO–T–228 (Chevy Chase, Md.: Operations Research Office, 1953), pp. 70–71.

[5] *Ibid.,* p. 71.

[6] *Ibid.,* p. 73.

[7] Adapted from Barton, *Paramilitary Warfare,* p. 74. This is a cumulative distribution for Korea, Malaya, and the Philippines. See pp. 15, 33, and 53 for tables for each country.

[8] *Ibid.,* p. 15.

[9] *Ibid.,* pp. 14–15

[10] *Ibid.,* p. 33.

[11] Adapted from Barton, *Paramilitary Warfare,* p. 33.

[12] *Ibid.,* p. 53.

[13] U.S. Information Service, *VC Battalion 261—Quantitative Profile* (Saigon: U.S. Information Service, March 1964).

[14] Barton, *Paramilitary Warfare,* pp. 74–75.

[15] Adapted from Barton, *Paramilitary Warfare,* p. 15.

[16] *Ibid.,* p. 32.

[17] George K. Tanham, *Communist Revolutionary Warfare* (New York: Praeger, 1961), p. 58.

[18] Alvin H. Scaff, *The Philippine Answer to Communism* (Stanford, Calif.: Stanford University Press, 1955), pp. 116–17.

[19] *Ibid.,* pp. 118–22.

[20] Tanham, *Communist Revolutionary Warfare,* p. 57.

[21] Barton, *Paramilitary Warfare,* p. 82.

[22] Scaff, *Philippine Answer,* pp. 116, 122.

[23] Tanham, *Communist Revolutionary Warfare,* p. 57.

[24] Scaff, *Philippine Answer,* p. 121.

[25] *Ibid.,* p. 118.

[26] Tanham, *Communist Revolutionary Warfare,* p. 57.

[27] Scaff, *Philippine Answer,* p. 119.

[28] S. A. Stouffer, *et al.*, *Studies in Social Psychology in World War II*, 2 vols. (Princeton, N.J.: Princeton University Press, 1949).

[29] Edward Shils and Morris Janowitz, "Cohesion and Disintegration in the Wehrmachts in World War II," *Public Opinion and Propaganda*, eds. D. Katz, D. Cartwright, S. Elderwold, and A. McLee (New York: Dryden, 1954).

[30] Scaff, *Philippine Answer*, pp. 121–22.

[31] Barton, *Paramilitary Warfare*, p. 82.

[32] Seymour Topping, "Portrait of Life with the Viet Cong: A Defector's Own Story," *The New York Times*, May 23, 1965, p. E–3.

[33] *Ibid.*

[34] Adapted from Republic of Vietnam, *Vietnam's Chieu-Hoi Policy* (Saigon: Psychological Warfare Directorate, Department of National Defense, June 1963), p. 28.

[35] *Ibid.*

[36] Molnar, *Undergrounds*, pp. 14–15.

[37] Adapted from Republic of Vietnam, *Chieu-Hoi Policy*.

[38] Seymour Topping, "Defection Plan of Saigon Lags," *The New York Times*, August 8, 1965.

[39] Topping, "Life with the Viet Cong," p. E–3.

[40] Andrew, R. Molnar, *Considerations for a Counterinsurgency Defection Program* (Washington, D.C.: Special Operations Research Office, 1965).

[41] Scaff, *Philippine Answer*, p. 112.

[42] *Ibid.*, pp. 123–24.

[43] All of the following material on Malaya was drawn from Lucian Pye, *Guerrilla Communism in Malaya* (Princeton, N.J.: Princeton University Press, 1956), pp. 324–38.

CHAPTER 4

IDEOLOGY AND GROUP BEHAVIOR

Common to most underground movements is an ideology, a set of inter-related beliefs, values, and norms. Ideologies are usually highly abstract and complex.[1] An ideology is more than a group of rationalizations and myths that justify the existence of a group; it can be used to manipulate and influence the behavior of the individuals within the group.[2]

In every society ideas, knowledge, lore, superstitions, myths, and legends are shared by its members. These are cultural beliefs. Associated with each belief are values—the "right" or "wrong" judgments that guide individual actions. This value code is reinforced through a system of rewards and punishments dispensed to members within the group. In this way, approved patterns of behavior, or "norms" are established.[3]

Human beings dislike ambiguity and uncertainty in their social and physical environment. Through generalized beliefs individuals seek to give meaning and organization to unexplained events. Common agreement on certain beliefs also enables individuals to operate collectively toward a desired goal. Leaders can interpret ambiguous situations in terms of the group's beliefs or ideology, translating abstract, ideological beliefs into specific, concrete situations in which actions are to be taken.[4]

Because beliefs and values are only distantly related to concrete action in daily life, an interpretive process is essential to derive specific rules of behavior. Commonly agreed-upon historical truths are used to justify the norms, values, and beliefs of the group. Significant events which occurred in distant times are given symbolic meanings, and a reinterpretation or "reification" of these events in the form of myths or legends supporting the group's purpose is developed. In doing this, the group may select certain concepts and adapt or distort them to justify specific forms of behavior; where existing concepts conflict with current activities, the group may deny that a particular concept is relevant in a particular case.[5]

Within any organization, there are reification sources whose role it is to apply official interpretations to significant changes in the social environment. An example of this is the Communist Party theoretician who modifies the official party line to fit world events.

In established groups, many beliefs are based upon authority; that is, since they are voiced by the leaders of the group, they are accepted as true. When a leader controls the dissemination of information to the members of an organization, he censors and approves various types of information. As a result, the group receives a restricted range of information, and group members tend to develop a set of common beliefs. Thus, in some cases, members need not be persuaded by argument, induced by reward, compelled by pressure, guided by past beliefs, or influenced by the opinions of other people; the restricted range of information to which they have access is sufficient to determine their beliefs.[6]

There is a constant drive among people to understand and validate certain beliefs. However, experiences are not always based on first-hand observations but upon second-hand accounts. Beliefs that cannot be validated empirically by events may be verified through a process called "consensual agreement," or agreement within the group. That is, each individual says to himself that whatever everyone else believes must be true. Members of a group tend first to seek a consensus within the group and then enforce the decision of the group. If subsequent events do not justify a group's beliefs or behavior, it may redefine the real world. A group tends to rationalize any situation and to blame external factors rather than internal group behavior.[7]

Within organizations, certain rules specify desirable behavior and the consequences of not conforming. The rules are enforced by organized rewards and punishments that are relevant to the objectives of the group. Normative standards are also enforced by surveillance of members.

INDIVIDUALS AND GROUP MEMBERSHIP

Group membership, such as in an underground, serves to satisfy several types of individual needs. It satisfies the need to "belong" and offers recognition and prestige. The member's status is enhanced and self-esteem is raised. A strong organization protects its members from external threats. It also gives opportunities to gain economic or political goals which could not otherwise be obtained.[8]

What others think of us, their praise or reproof for our actions, affects our self-esteem. Threats to our self-esteem can motivate us to perform poorly or well. Thus, group assessments of individual performance can exert a strong influence on our behavior.[9]

The individual carries out ideas to which he thinks the group aspires and is either encouraged or intimated by how the group responds to his actions. Further, an individual's "level of aspiration" is influenced by the standards of his group and his culture, and his group can influence him to raise or lower that level.

People see things—other people, objects, and events—not in isolation but within a frame of reference (the standard or yardstick by which an individual evaluates new information). To understand what a person sees when he looks at an object or event, we often need to know the properties of the frame of reference to which he relates the perceived information. Identical events can be given different meanings, depending upon individual frames of reference. If a landowner distributes money among his peasants, some may be irritated, feeling that of all the wealth he "stole" from the peasants he is returning only a small part, while others interpret the act as a sign of generosity.

A man in certain circumstances and situations can be led to make decisions and take actions which might be against his better judgment under a totally new frame of reference in order to interpret a particular event, or when his mental context is so rigidly fixed that all events are rationalized within it.[10] In either case, group pressure can lead an individual to take action consistent with group objectives.

FACTORS RELATED TO GROUP INFLUENCE

In his efforts to make his environment meaningful, man distorts, emphasizes, and suppresses the information he receives. Further, he perceives only a limited number of those things to which he is exposed, and tends to listen to those which interest him most. Selective exposure to environment is furthered by membership in occupational and other groups.

Human perception is also limited by the tendency to concentrate on a few immediate alternatives when making a decision. Instead of developing other, or more logical, alternatives, man tends to take the best of those available to him at the moment.[11] This tendency is probably one reason why situational factors are so important in recruitment and defection of insurgents.

When events contradict normal expectations, the need to understand and explain the frustration frequently leads to a distortion of facts. When Stalin concluded a nonaggression pact with Hitler, until that time the foremost enemy, Communist members had to rationalize the situation and remain loyal.

An individual conforms to group norms for many reasons. He may conform out of habit, he may anticipate group-administered rewards, such as promotion or group approval, or he may be directed through the use of group disapproval or sanctions.[12] Group signs of disapproval ranging from mild disapprobation to utter condemnation are immediately recognized by members. For slight deviations from group norms, the group may withdraw signs of approval rather than offer reproach. Such psychological sanctions depend for their effectiveness chiefly on the value which the individual places on his status in the group. While great loss of status seldom results from a single act, serious or persistent deviations may lead to partial loss of status, in which recognition of the individual as a "member in good standing" may be temporarily withheld.

The leader's authority is often sufficient to maintain social control. Persuasion may be used to present a particular judgment in such a way that the individual members see the value of accepting it in place of their own judgment. They remain free to decide how and in what way they will act. The group can also be manipulated by a calculated presentation of facts, or the unit consensus may be represented as the total group when in fact it is only a small part of it. Coercion may be used, but since it may create a high degree of alienation, only as a last resort; even then, the threat of physical punishment is more often used than the actuality.[13]

Expulsion from the group is another form of punishment. This may have extreme effects upon the individual, especially if he is dependent upon members in line.

A group's code of normative behavior is largely implicit; members know what is right and wrong, what can and cannot be done, but find it difficult to express the code in words. They cannot define a role but can say "this is the way it is done."[15]

Within any organization, the individual has a status position. Certain things are expected of him. He plays his role and expects others to act toward him in certain ways. Status roles involve a set of clearly defined and rigidly maintained rights, including such "status symbols" as special uniforms and insignia and deference from others.[16]

Rituals or ceremonies may be developed to bring about normative behavior as well as to create a feeling of belonging. They may involve signs or symbols signifying membership in the organization or ceremonial entrance and initiation rituals. Most underground organizations have a formal initiation and pledging ceremony which attempts to impress on the new member the value of membership in the group as well as the desirability of conforming to the group's point of view and group norms. Candidacy for a group may involve considerable pretesting. A candidate may be invited to participate on a guest basis, then may face an election, and finally be inducted into the main group. Such procedures are not always functional; sometimes they are purely ritualistic and valued in themselves.[17]

There are several factors which will determine how much influence the group exercises over the individual through group pressures and norms. One factor is the size of the group. The smaller the group, the more effectively control is exerted over an individual. Other things being equal, the control exercised by the group is in inverse proportion to its size.

The frequency with which a group meets also affects the relationships of its members—the more often it meets, the more intimate the relationships within the group. The ability of the group to control the behavior of an individual is directly related to the length of time that the group has existed and the frequency of its members' contacts with each other.

In highly structured groups, whose members' relationships and duties are clearly defined and in which there are generally recognized norms, more control is exercised over members than in loosely structured groups.[18]

FACTORS RELATED TO CONFORMITY

Factual matters and personal preferences are resistant to change, while political ideologies, social attitudes, and expressions of opinion are susceptible to pressures toward conformity.[19] Ambiguous topics are more changeable than clear-cut ones and suggestibility increases when it is difficult to check the accuracy of one's response. Knowledge of another person's

---

* In Greece, during World War II, a member of the National Liberation Front (EAM) was expelled and threatened with physical harm. He felt compelled to join the rival underground group to get protection against his former associates.[14]

94

response in a similar situation also increases suggestibility, especially when influence is exerted by a person of higher status than oneself. Suggestibility is also greater under pressures from friends or acquaintances than from strangers. Conformity grows progressively with the size of the group. However, influences toward conformity decrease when other group members are not in unanimous agreement. When interaction among group members is increased or a permissive group-centered approach is employed, susceptibility to conformity is increased. Susceptibility to group pressures is greater when rewards are given for "successful" performance and penalties for "mistakes," and among group members who have shared success.

An individual who makes a definite commitment and is then subjected to pressure tends to resist and maintain his position strongly. This is especially true if the commitment has been made in public. This is one reason why insurgents require formal oaths for new members of their movement and why they insist upon symbolic acts of public commitment in exercising population control.

If the individual displays anxiety before a pressure situation appears, he tends to be more susceptible to suggestion. Also, young people consistently give in to social pressure more readily than older persons. This is probably why young people are more readily coerced into joining insurgent movements. Loss of sleep, too, tends to make an individual more susceptible to pressures.

Tendencies toward conformity increase when the views presented to a person appear to him to be only slightly different from what he believes to be his own convictions. Resistance is minimized when the new views are presented gradually in small steps. This is reflected in underground recruitment techniques.

Conversion, like conformity, is highest among persons who are uncertain about factual matters. The degree of conversion is limited by how intensely a man held his initial views. The less intensely he held them, the easier it is to convert him. The longer an individual resists, the longer he will stick to a new position once he has altered his views. It has been found that behavior which was altered by social pressure generally persists over time.

UNDERGROUND ORGANIZATION PROCESSES

Underground movements can be described as "normative coercive" organizations. They are normative in that they appeal to people by offering to satisfy certain goals and to provide rewards, prestige, and esteem. However, coercive power is also applied through the threat of deprivation of certain satisfactions or the application of physical sanctions such as pain, deformity, or death.[20]

Although an individual may be persuaded, coerced, tricked, or forced to join the movement, his goals and desires change as he stays with the organization. Recruitment is only the initial phase of involvement. Indoctrination brings about a socialization of the individual, and his experiences in participation with members of the movement change his attitude and eventually his goals.[21]

INSULATION AND ABSORPTION

During the indoctrination period, the aim is to have the individual internalize the values of the organization. Total control is achieved through insulation and absorption. Through ideology the individual is insulated and given a separate moral and intellectual world within which to think and operate; all events are interpreted within the context of ideology.[22] The conspiratorial atmosphere, with an emphasis on illegal work, starts a process of disintegration of normal moral principles and a reduction of inhibitions which hampers an individual's actions and manipulability. All of his time is absorbed by organizational activities—meetings, demonstrations, distribution of literature, and recruitment. This constant activity gives the individual's life an apparent meaning and removes him from outside interests and contacts.[23]

Indoctrination and education tend to reinforce an individual's loyalty to the underground organization and to immerse him in the movement. The individual is disciplined and schooled to think in terms of how individual actions help or hinder the organization, not how they suit personal goals.[24]

When an individual joins an organization, the number of decisions and alternatives available to him decreases. That is, he devotes most of his time to organizational activity and therefore limits his outside interests. As an individual reduces the number of personal relationships with others, he tends to internalize the rules of the organization and the less he searches for alternate forms of behavior. Small, closely knit cohesive groups are highly predictable in behavior. This rigidity increases the extent to which the group goals are perceived to be shared by all the members of the group, and thus its esprit de corps. In this manner individuals within the group protect themselves from outside pressures.[25]

The internalization process is complete when the group member maintains his conduct without such enforcing agents as surveillance or direct threat of punishment and when he performs his duties for their own sake. As the individual builds up institutional habits and internalizes a code of conduct, he is less likely to leave the organization.[26] The smaller the group, the greater the individual's involvement and compatibility. The larger the organization, the greater the chance of conflict. The more extensive his participation in group activities, the more likely the individual is to develop loyalty and moral involvement and finally a commitment to the general goals of the organization.[27]

Frequent assignments and a high degree of activity also have a useful side effect, providing the individual with an "invulnerability concept." He becomes so engrossed in his work that he loses any fear of harm coming to him. While aware that others have been caught, he is so busy with his daily routine that he unconsciously considers himself invulnerable.

Joining an underground movement is quite different from joining an ordinary political group. One ex-Communist says that it is not like joining a political party but like joining a church. It is a way of life.[28] Another former undergrounder says:

> A faith is not acquired by reasoning. One does not fall in love with a woman, or enter the womb of a church, as the result of logical persuasion. Reason may defend an act of faith—but only after the act has been committed, and the man committed to the act. Persuasion may play a part in a man's conversion; but only the part of bringing to its full and conscious climax a process which has been maturing in regions where no persuasion can penetrate.[29]

The act of commitment in insurgent organizations is uniformly an oath-taking process. The individual performs some symbolic, overt act which demonstrates that he is willing to accept the rules of the organization and abide by its sanctions if he does not conform. Once committed, the individual reorganizes his frame of reference and the way he views the world to conform to his commitments.

There are several major mechanisms for keeping members cooperative and working in unison. An individual rises to leadership positions by being highly active and then assuming a full-time position within the organization. Those individuals who have special qualifications but lack essential disciplinary characteristics can be put in special positions through the process of cooptation: They are put on certain committees and participate in some organizational activities without following the rules of ordinary membership. Individuals outside of the organization may also be brought in to support the organizational goals through cooptation. An individual who does not agree with the subversive group's main goals or activities may be brought to support the organization by giving him assistance in attaining limited goals which the individual favors within the community. This happens frequently in front groups. An individual may favor disarmament or be opposed to the government for specific reasons; the subversive group sponsors and supports him, obtaining his loyalty in return.[30]

The atmosphere of the underground has been described as—
> ... a paradoxical atmosphere—a blend of fraternal comradeship and mutual distrust. Its motto might have been: Love your comrade but don't trust him an inch—both in your own interest, for he may betray you; and in his, because the less he is tempted to betray, the better for him. This, of course, is true of every underground movement; and it was so much taken for granted that nobody seemed to realize the gradual transformation of character and of human relationships which a long Party career infallibly produced.[31]

COMMUNIST FACTORS IN ORGANIZATION PROCESSES

Communist organizations are characterized by five major organizational factors: (1) ideology, (2) democratic centralism, (3) criticism and self-criticism techniques, (4) the committee system, and (5) cell structure. The effectiveness and use of these organizational techniques rest on some basic principles of social organization.

izations. They are not primarily designed to teach specific or detailed ideological content. The organization coins value terms—such as "deviationist" and "personality cult"—which to the outsider have no real semantic value but within the organization carry positive or negative connotations and indicate to the membership those things which the leadership favors or does not favor. Through constant indoctrination, the value systems these terms represent are inculcated into the membership and normative patterns of behavior are developed.

Individuals tend to abide by a decision as long as they are permitted to voice their opinions, notwithstanding the outcome. Through the principle of democratic centralism, the Communists have capitalized upon this common social phenomenon. Members are seemingly permitted to participate in the decision-making process even though the leadership fully controls the structure. Although the Communist organization does not allow free elections, and even though decisions are largely made in advance, the membership does discuss and criticize issues before they are guided to the "correct" position. Further, if the individual has supported a measure even superficially, he is more committed to carrying out required action.

All cells within the party hold criticism and self-criticism sessions in which each member must criticize others' activities as well as his own. These sessions permit the leadership to better understand the individual member's capabilities and problems and provide social pressure to reinforce normative behavior. The group discussion provides consensual validation of group beliefs, and the individual can justify his behavior in his own mind because the rest of the group approves of it. The leader also can use the self-criticism session to praise members. However, precautions are taken to prevent attacks on the goals of the party itself; criticism is directed toward improving means to further the ends of the party. Furthermore, each individual must state his plan to correct his defects; the group thus requires him to raise his level of aspiration in performing his duties.

The committee system provides ample opportunity for everyone to participate in leadership roles. Members learn to be leaders in this way, and in the process the group satisfies the power desires of its members.

The small size of the cell makes it a more cohesive group and tends to reduce inner frictions. As stated earlier, the smaller the group, the more effective its control over its members. Where the party is illegal, and where such control is especially important, cells average four to six members.

In summary, it may be seen that the Communist Party has evolved value and norm systems as well as organizational mechanisms which create a high degree of cohesiveness in its operations. Furthermore, the techniques seem to be effective in providing informational feedback to the leadership. The criticism and self-criticism sessions apply social pressure to reinforce the behavior patterns acquired through indoctrination; they also increase levels of aspiration and commitment.

FOOTNOTES

[1] David Krech, Richard S. Crutchfield, and Egerton L. Ballachey, *Individual in Society* (New York: McGraw-Hill, 1962), pp. 402; 42.

[2] Richard T. LaPiere, *A Theory of Social Control* (New York: McGraw-Hill, 1954), p. 240.

[3] Krech, Crutchfield, and Ballachey, *Individual in Society*, pp. 349–50.

[4] Neil J. Smelser, *Theory of Collective Behavior* (New York: The Free Press of Glencoe, 1963), pp. 81–82.

[5] LaPiere, *Theory of Social Control*, pp. 260–66.

[6] Krech, Crutchfield, and Ballachey, *Individual in Society*, pp. 402–403.

[7] *Ibid.*, p. 404.

[8] *Ibid.*, p. 394.

[9] Wilbur Schramm, *et al.*, *The Nature of Psychological Warfare*, ORO–T–214 (Chevy Chase, Md.: Operations Research Office, 5 January 1953), pp. 87–93.

[10] Hadley Cantril, *The Psychology of Social Movements* (Science Editions [paperback], 1963), (New York: John Wiley and Sons, 1941), p. 65.

[11] James G. March and Herbert A. Simon, *Organizations* (New York: John Wiley, and Sons, 1961), pp. 169–71, 190.

[12] LaPiere, *Theory of Social Control*, p. 239.

[13] M. Haire (ed.), *Modern Organizational Theory* (New York: John Wiley, 1959), pp. 107–108.

[14] Interview with former member of the Greek EAM underground.

[15] LaPiere, *Theory of Social Control*, pp. 188–212.

[16] *Ibid.*, p. 80.

[17] *Ibid.*, p. 209–14.

[18] *Ibid.*, pp. 101–106.

[19] Robert R. Blake and Jane S. Mouton, "The Experimental Investigation of Interpersonal Influence," *The Manipulation of Human Behavior*, eds. Albert D. Bidderman and Herbert Zimmer (New York: John Wiley, 1961), pp. 216–76.

[20] Amitai Etzioni, *A Comparative Analysis of Complex Organizations* (Glencoe, Ill.: The Free Press, 1961), pp. 56–59.

[21] Etzioni, *Comparative Analysis*, p. 152; Philip Selznick, *The Organizational Weapon: A Study of Bolshevik Strategy and Tactics* (Glencoe, Ill.: The Free Press, 1960), p. 24.

[22] Selznick, *The Organizational Weapon*, p. 25.

[23] *Ibid.*, pp. 26–27.

[24] *Ibid.*, p. 47.

[25] March and Simon, *Organizations*, p. 39.

[26] *Ibid.*, p. 105.

[27] Etzioni, *Comparative Analysis*, p. 287.

[28] A. Rossi, *A Communist Party in Action* (New Haven, Conn.: Yale University Press, 1949), p. 15.

[29] Arthur Koestler, "The Initiates," *The God That Failed*, ed. Richard Crossman (New York: Harper and Brothers, 1949), p. 15.

[30] Etzioni, *Comparative Analysis*, p. 103.

[31] Koestler, "The Initiates," pp. 29–30.

CHAPTER 5

CLANDESTINE AND COVERT BEHAVIOR

A former underground leader has suggested that while it is difficult to completely escape from modern scientific surveillance methods, there are many ways to mislead the surveillants. The underground member, wishing to minimize risk and chance factors, attempts to be as inconspicuous as possible and refrains from activities which might bring attention or notoriety.[1] He strives to make his activities conform with the normal behavior and everyday activities of the society in which he lives. By appearing conventional and inconspicuous, he makes it difficult for the security force to detect, identify, or locate him. Besides making himself inconspicuous, the underground member avoids materials or contacts that might give him away. Subversives keep a minimum of records and contact other agents only when essential. Without physical evidence, signed confessions, or defectors who accuse others, it is difficult to link an individual to a subversive organization. Contact and communications between agents is considered the most critical phase of subversive operations.

Even when a subversive is identified and his activities known, the practice of clandestine and covert behavior makes it exceedingly difficult for security forces to locate him among a country's millions of citizens.

DEFINITIONS OF CLANDESTINE AND COVERT BEHAVIOR

Both clandestine and covert operations are secret, but in different ways. Clandestine operations are those whose existence is concealed, because the mere observation of them betrays their illegal and subversive nature. Secrecy depends upon skill in hiding the operation and rendering it invisible. For example, weapons might be manufactured in some rural redoubt, out of view and hidden from the eyes of the security forces. Covert operations are usually legal activities that serve as a cover for their concealed, illegal sponsorship. In short, clandestine behavior is hidden from view, while covert behavior is disguised to conceal its subversive character.[2]

A classic example of a covert operation appeared in Italy at the end of World War II. The Mossad was the Jewish underground organization in charge of the movement of Jews from Central Europe to Mediterranean ports. In 1946, the Mossad unit operating in Italy found it impossible to move on roads and obtain fuel and spare parts without special licenses and permits. The unit had acquired 40 British Army vehicles, but a major problem was how to maintain and fuel the vehicles without arousing the suspicion of the British military police. To provide the needed cover, the Mossad created and staffed an imitation British Army installation of their own on land "requisitioned" near a town. Since many of the underground had served in the British Army, they were able to duplicate the authentic military atmosphere. The camp was complete with badges, insignia, and notices. It had identification numbers in accordance with British practice, and local laborers were employed. With forged company papers, work tickets, and requisition papers, they took over a large courtyard and garage in the center of Milan. They provided themselves with everything that the Army should have—signboards, official documents, papers, guards, motor pool, and so on. The whole installation operated without suspicion from the Italian civilians it employed or even from the genuine British soldiers who periodically brought their jeeps into the motor pool for gasoline. In this manner, the Mossad provided both provisions and fuel for the motor vehicles that transported many thousands of refugees. The camp functioned for two years without arousing suspicion. It attained the reputation among the British camps in the zone as being a proper, well-disciplined camp, which would "not issue even a drop of petrol without orders." [3]

TECHNIQUES OF CLANDESTINE BEHAVIOR

The techniques of clandestine and covert behavior utilized by undergrounds are multifarious. They involve organizational devices, patterns of communication, and rigid security procedures.[*]

ORGANIZATION

Subversive organizations try to distribute cells and units over widely separated geographic areas and among different ethnic and social groups. In this way, the government security forces can be extended so that they cannot concentrate on any single area or social group. The Malayan Communist Party, made up almost entirely of indigenous Chinese, was easy prey for the security forces. On the other hand, the FLN in Algeria was composed mostly of Muslims, who constituted 90 percent of the population, and its cells were widely distributed throughout the country.

The specific geographic location of the cell can minimize the chance of being detected. During the Palestine insurgency the Jews set up a major transportation center less than 100 yards from the headquarters of the British forces in Tel Aviv but aroused no suspicion. However, it was also near a powerplant and a central bus station and amid many garages and auto workshops where day-and-night traffic was normal.[4] Similarly, in Algeria, Yassef Saadi, a political-military commissar of the FLN, took advantage of the hustle of a busy spot, and set up his offices only 200 yards from the office of the Army Commandant of the Algiers Section.[5]

Individual cell members are instructed to seek no more information than is required to perform their tasks.[6] The cell members must go through an intermediary or through a mail-drop in order to get in touch with the cell leader, whose identity and location are unknown to them. Liaison between echelons is so regulated that a captured member cannot lead his captors to

---

[*] For a brief summary of the rules of clandestine and covert behavior practiced by five underground organizations during World War II, see appendix C.

the next highest official with whom he regularly conducts business. All contacts with higher echelons are prearranged through intermediaries, and the higher official sets the time and place for meeting. If a cell member is captured, the chain between the cell and the leader is broken, thus cutting off the cell from the organization and protecting the underground organization from compromise. This fail-safe principle is found in almost all underground organizations and operations.[7]

Records are kept to a minimum; wherever possible, information is memorized rather than recorded. Messages are coded in some manner before being written down. Cover names are used in order to protect the identities of the people. Individual members are instructed not to keep any written messages or diaries.[8] False units are established and communications and records created for them in order to confuse the security forces should they acquire or capture any organizational records.[9]

COMMUNICATIONS

Communicating with another underground member is the most dangerous activity in clandestine or covert operations.[10] A cardinal rule in underground operations is that agents should be seen together in public only when absolutely necessary. They usually work through an intermediary who meets each agent separately and conveys messages back and forth. The use of couriers is probably the safest means of communication and transmission of information between various agents. Ninety percent of all communications in the Philippines insurgency and in Korea involved the use of couriers. Couriers generally are children, women, or aged men, who can move about without drawing attention to themselves. This was true in Italy, in Poland, and in Belgium. Preferably, the courier should travel as a natural part of his job, and such people as taxi drivers and traveling vendors make good couriers.[11]

Another means of transmitting information between agents is the maildrop. The underground agents come by prearranged schedule, one at a time, to a particular location where a message is left. The location of the mail-drop must be a natural and safe place, such as an old log in a park, where strangers will not accidentally pick up the message. Usually, a reserve drop is designated in the event the first one is unusable. The communication should not be left in the mail-drop for long. Signals are prearranged for each drop, usually at a different location, so that the agent will know when the drop is full or when it is empty. In this way he need not go there and perhaps arouse suspicion only to find that there is no communication.[12] Ideal locations for mail-drops or for the alerting signals are places such as telephone booths or washrooms where an individual commonly goes alone without suspicion.[13]

The telephone is seldom used; agents are usually forbidden to call each other directly. If telephones are used in an emergency, the individual goes to a pay station and uses a prearranged code. Neither is the open mail used often. If mail must be received from abroad, it is sent to a cover address, which may be that of a person who has frequent visitors, such as a merchant. Mail is never delivered to an agent's house. He goes to the post office to get it so no letter can be traced through the mails directly to an agent.[14]

Strict rules for meetings are observed. Underground members are careful not to use the same meeting place too frequently. Before the meeting, the family at whose house the meeting will be held is checked to be sure that they are thoroughly reliable. Someone within the family is assigned to answer the door in case an outsider knocks and to serve as a lookout. Times of arrival and departure are staggered to avoid attracting attention to large groups coming or going.

Once at the meeting, explanations or cover stories are arranged among the members in case the meeting should be broken up by security forces and the group interrogated separately. A reason for the gathering is established, such as "getting together to play cards." Others would be such occasions as birthdays, anniversaries, or weddings. No more documents are taken or carried to the meeting than are absolutely necessary. Note-taking is not permitted; the individual must use his memory. After the meeting, a rearguard checks to be sure that no materials have been forgotten.[15]

When meeting in public, visual identification marks and passwords are usually used as recognition signals. Generally, passwords are innocent-sounding, so that if the wrong person is approached or the exchange is overheard by bystanders, it will not be interpreted with suspicion. The password may ask directions or make similar innocent requests. Visual identification marks include the wearing of unique combinations of clothes or the carrying of specified objects. For any meeting between agents, should contact be missed on the first try, a second place is prearranged for ten minutes later.[16]

There are many ways in which an individual takes precautions to insure against being followed to a rendezvous with another agent. When driving to make a contact, an individual can alter his speed, enter intersections on a yellow light, or turn a corner and stop abruptly. Or he may use the switch-point technique: he drives to a particular location in a "drop car," gets out of the car, and walks across a parking lot or into a department store to another location where he is picked up by a second car (the "pick up"), is driven to another place to be picked up by a third vehicle, and then is taken to his destination.[17]

When traveling on foot, an agent attempts to leave a subway at the last minute or to enter a hotel or bus terminal at one door and leave by another. He can also use a store window along a main street as a mirror to see if anyone is following him.[18]

SECURITY

Recruits are usually not accepted until their family life, jobs, political activities, and close associates have been investigated. Most undergrounds also require a probationary membership period. The individual is assigned limited tasks and his contacts with other members of the cell are restricted.

Even in guerrilla units, the recruit is given no assignments that would bring him in contact with outsiders. The new member goes through an indoctrination period and is given tests in order to determine how he thinks and what he feels on particular issues. In addition, disciplinary squads are used to inspect the belongings of individuals on occasion to find out if they have violated any security rules.[19]

Most underground movements require new members to take a loyalty oath designed to commit them to the organization and impress upon them the seriousness of the job. Such oaths usually require the individual to accept all missions and obey all orders on penalty of death. Sometimes there are lesser penalties for lesser violations of security rules.[20]

HUMAN FACTORS IN CLANDESTINE AND COVERT BEHAVIOR

The most serious danger in clandestine operations comes not from spies or infiltrators but from the inadequacy of the human beings who compose the underground. One of the most critical areas of underground work is the teaching of members to maintain silence. Normal curiosity leads members to find out more information than they should know. A second difficulty is that people want to talk about their accomplishments to someone; it is through idle talk and unguarded conversations that most clandestine organizations are compromised. To guard against this, undergrounds stress discipline. At the same time, they try to develop a sense of discretion among the members so that "adherence to the rules" won't stifle initiative. Maximum observation of rules can lead to passivity and inaction, so members must be willing to waive any rule and attempt any action which circumstances demand, if the end seems worth the risk.[21]

Within any society there are customs and norms by which people abide without question. By capitalizing upon these customs the underground can carry out many of its activities in a normal manner and under good cover. For example, one woman courier carried a message concealed among half a dozen eggs in a bag; the guard, afraid of breaking the eggs, did not inspect the bag closely.[22] The role of women and children in most societies is a protected one and they are usually beyond suspicion, which is why they are so frequently used as couriers.

Certain social roles can be assumed to avoid interrogation. Pretending to be insane, deaf-and-dumb, or sick—e.g., having a toothache—is effective because such roles are generally accepted without question.[23] Similarly, certain locations are unlikely to be investigated. One courier who could not find lodgings sought out the redlight district of a major city; he assumed that the underworld had agreements with the police and he could be safe for a short time in these quarters.

In populations where there are many subcultures, a knowledge of normative behavior is critical. During the Arab-Israeli campaign, Arabic-speaking Israeli raiders disguised themselves as Arab military personnel, police, tribesmen, or pilgrims to carry out sabotage missions. A special unit of dark-skinned Jews from Yemen, Iraq, Syria, and Morocco was formed. This "Black Regiment" had many successful exploits. However, in some cases their operations failed because of human error. For example, not knowing that only Arab officers—not the rank-and-file—use handkerchiefs and toilet paper jeopardized missions. A Mecca-bound Muslim pilgrim never relieves himself facing east, for this is the direction of the holy city and is reserved for prayer. Further, it is a grave offense to clean one's nose with a finger of the left hand; this hand is reserved for the lavatory. Other missions failed because the raiders smoked Israeli-made cigarettes and dropped butts along the beach after clandestine landings—Arabs not only did not smoke Israeli cigarettes but seldom, if ever, threw cigarette butts away. These cultural factors compromised many of the disguises of the raiders.[24]

Another example of habit being relied upon for use as cover occurred in Palestine. An illegal cargo was covered with a tarpaulin and a layer of fresh manure. The police disliked searching such a load too closely and the cargo got through police inspection without being stopped.[25]

An individual who has a preexisting frame of reference reacts immediately to new events, without reflection. When a person's mental context is thus fixed, he possesses what has been described as a "will to believe."[26] This factor may be capitalized upon by the underground by knowing the mentality and habits of the police. For example, in Palestine trucks with insignias of well-known transport companies were used to transport illegal cargo. A "policeman" preceded the truck, which appeared to be moving from an established factory.[27] In addition, diversionary efforts were made: a man would strike a policeman just as the truck was about to pass. The resulting commotion permitted the truck to go by unnoticed. Another device used to avoid roadblocks was to have trucks join military convoys. It was necessary, however, to call ahead and include in the convoy two or three trucks from "another unit." The unit was usually expressed in some abbreviation which would be unquestioned by the men at roadblocks.

Trademarks of well-known products have been put on illicit equipment. Capsules for explosives have been marked "Bayer" (like the aspirin). In this manner, suggestion and the use of normal everyday patterns of behavior have concealed illegal activities.[28]

Within any society, there are symbols associated with certain roles. Individuals can be recognized as important, but not specifically identified, by the uniform or clothes worn. In one operation, an Israeli reconnaissance squad requisitioned a set of fancy dress uniforms from the Hebrew National Opera in Tel Aviv. The commanding officer put on a 19th-century, Imperial Austro-Hungarian Hussar uniform with gold shoulder boards and glittering buttons. In a white jeep, the squad calmly drove across the Arab lines and introduced themselves to the sentries at the first control checkpoint as United Nations military observers from Luxembourg. They cruised up and down the enemy lines for a day and even lunched with an Arab colonel.[29]

Sometimes an underground establishes certain innocent patterns of behavior that it later capitalizes on. For example, in Palestine an isolated seashore police post was penetrated by a young boy and girl working for the Haganah. They went swimming daily, and, on leaving the bathing area, walked directly past the police guards. The police became accustomed to seeing the couple, and even the dog at the police station got to know them and stopped barking. In this manner, information was obtained for a raid.[30] By conditioning the guards to an apparently innocent pattern of behavior, the underground was able to take advantage of opportunities for gathering intelligence.

FOOTNOTES

[1] Jon B. Jansen and Stefan Weyl, *The Silent War* (New York: Lippincott, 1943), pp. 151–52.

[2] Christopher Felix, *A Short Course in the Secret War* (New York: E. P. Dutton, 1963), p. 28.

[3] Jon and David Kimche, *The Secret Roads* (New York: Farrar, Straus and Cudahy, 1955), pp. 118–20; Gershon Rivlin, "Some Aspects of Clandestine Arms Production and Arms Smuggling," *Inspection for Disarmament*, ed. Seymour Melman (New York: Columbia University Press, 1958), pp. 197–98.

[4] Rivlin, "Clandestine Arms," p. 195.

[5] Roger Trinquier, *Modern Warfare* (New York: Praeger, 1964), p. 15.

[6] Hans J. Reichhardt, "New Beginnings: A Contribution to the History of the Resistance of the Labor Movement Against National Socialism," (unpublished mimeographed manuscript, c. 1961); see also David J. Dallin, *Soviet Espionage* (New Haven, Conn.: Yale University Press, 1955), p. 465.

[7] Dallin, *Soviet Espionage*, pp. 463–65; see also J. Edgar Hoover, *Masters of Deceit* (New York: Henry Holt, 1958), p. 280.

[8] Dallin, *Soviet Espionage*, p. 9.

[9] George K. Tanham, "The Belgian Underground Movement 1940–1944" (unpublished Ph.D. dissertation, Stanford University, 1951).

[10] Felix, *The Secret War*, p. 137.

[11] Jan Karski, *The Story of a Secret State* (London: Hodder and Stoughton, 1945), pp. 229–35; Dallin, *Soviet Espionage*, p. 9; Fred H. Barton, *Salient Operational Aspects of Paramilitary Warfare in Three Asian Areas*, ORO-T-228 (Chevy Chase, Md.: Operations Research Office, 1953), p. 24; L. Valiani, *Tutte Le Strade Conducono a Roma (All Roads Lead to Rome)* (Firenze: Tipocalcografia Classica, 1947), p. 158; T. Bor-Komorowski, *The Secret Army* (London: Victor Gollancz, 1950), pp. 50–60; Tanham, "Belgian Underground," pp. 147–49.

[12] Pawel Monat and John Dille, *Spy in the U.S.* (New York: Harper and Row, 1961), pp. 63–67.

[13] *Ibid.*, pp. 79–80.

[14] Dallin, *Soviet Espionage*, p. 9; Hoover, *Masters of Deceit*, p. 279; Tanham, "Belgian Underground," p. 147; Jansen and Weyl, *Silent War*, p. 113; and Bor-Komorowski, *Secret Army*, p. 59.

[15] Hoover, *Masters of Deceit*, pp. 282–83; Jansen and Weyl, *Silent War*, pp. 112, 153.

[16] *Ibid.*

[17] Hoover, *Masters of Deceit*, pp. 283–85.

[18] *Ibid.*, p. 284.

[19] Barton, *Paramilitary Warfare*, pp. 28–29; Philip Selznick, *The Organizational Weapon: A Study of Bolshevik Strategy and Tactics* (Glencoe, Ill.: The Free Press, 1960), p. 19; Reichhardt, "New Beginnings"; and Hoover, *Masters of Deceit*, p. 279.

[20] Rivlin, "Clandestine Arms," pp. 191–92; Tanham, "Belgian Underground," p. 178; Andrew R. Molnar, et al., *Undergrounds in Insurgent, Revolutionary and Resistance*

Warfare (Washington, D.C.: Special Operations Research Office, 1963), p. 79; and Ronald Seth, *The Undaunted: The Story of the Resistance in Western Europe* (New York: Philosophical Library, 1956), p. 259.

[21] Jansen and Weyl, *Silent War*, pp. 155 and 180–81.

[22] *Ibid.*, p. 157.

[23] Molnar, *Undergrounds*, p. 122.

[24] Leo Heiman, "All's Fair . . . ," *Marine Corps Gazette*, XLVIII (June 1964), pp. 37–40.

[25] Rivlin, "Clandestine Arms," p. 195.

[26] Hadley Cantril, *The Psychology of Social Movements* (Science Editions, 1963) (New York: John Wiley and Sons, 1941), p. 65.

[27] Rivlin, "Clandestine Arms," p. 196.

[28] *Ibid.*, p. 197.

[29] Heiman, "All's Fair . . . ," pp. 37–40.

[30] Rivlin, "Clandestine Arms," p. 199.

PART III

UNDERGROUND ADMINISTRATIVE OPERATIONS

INTRODUCTION

To develop an effective organization, undergrounds must perform certain basic administrative activities, such as recruiting qualified and loyal personnel, indoctrinating and training the membership, and obtaining financial support. In all of these activities a number of critical human factors are involved.

For instance, people must be recruited in such a manner that those who refuse to join do not later inform on the recruiters or expose the movement. Undergrounds must also devise means to persuade and deeply commit those who join for a whim. Training processes must be structured so that recruits are steeled to carry out dangerous assignments, yet remain loyal for long and stressful periods of time. In order to finance an insurgency, a regular supply of funds is required. Because tax collection or other imposition of financial burdens on a populace is unpopular even under normal circumstances, undergrounds must devise ways to get large sums of money while keeping the voluntary support and protection of the people.

The following three chapters discuss these three important administrative operations: recruitment, education and training, and finance. The performance of these administrative activities varies widely and is contingent upon the situation and the effectiveness of the security forces. There is no one best way to perform them, and so several approaches are described for each operation. For example, both selective and mass recruitment methods are discussed. In the chapter on education and training, topics and documents which have been considered important enough to be included in various training programs are reviewed. The chapter on finance attempts to bring together the little that has been written on collection techniques and places a special emphasis on human factors considerations.

CHAPTER 6

RECRUITMENT

The types of people recruited into an underground movement and how they are recruited depend largely upon the movement's stage of development. At first, primary attention is given to the development of a carefully chosen, well-disciplined cadre. Later, greater emphasis is given to developing mass support.

First the recruiter identifies talented people with grievances, surreptitiously tests each individual's loyalty to the government, and then through a process of gradual commitment leads the recruit into illegal and underground work. Recruits are seldom placed in positions where they must immediately decide to join or not to join the underground. Instead, through a series of seemingly innocent or slightly illegal acts which, when viewed by an outsider, appear subversive or illegal, the recruit is led to believe that an overt commitment to the underground is his only alternative. He suddenly finds himself in a position where to betray the underground he must also incriminate himself: if he does not join, the underground may tell the police about the illegal acts he has been enticed to commit.

Appeals to recruits are usually based upon the assumption that everyone has grievances, temptations, and vulnerabilities. It is the recruiter's task to uncover, crystallize, and exploit the right combination of these personal and situational factors. Appeals to ego, power, or recognition are strong factors in attracting individuals to a movement. Such rewards and profits are devices used to entice individuals and to keep them involved, whereas techniques of social pressure and threats of social sanction are used to obtain final commitment. In short, the recruit is attracted by making appeals which coincide with his value system, yet which lead him from lesser acts for profit to final acts of commitment.

Undergrounds seldom rely solely upon the good intentions of recruits. Typically, they avoid leaving anything to chance. Because reasons such as patriotism, social justice, or personal grievances may or may not be sufficient for attracting or sustaining recruits through the long dangerous struggle of protracted war, coercive measures are also implied, or even applied, by the underground. Incriminating evidence may be kept to insure that recruits do not defect. In the recruit's oath, a death penalty is usually the price for betrayal of organizational secrets or defection, and special terror squads are retained especially to carry out the penalty.

During mass recruitment the recruiter realizes that while many people will volunteer, others must be persuaded and coerced into joining. For those who are likely to volunteer (perhaps because of personal grievances), the recruiter finds that listening is a technique as important as persuasive argument. Professional recruiters also recognize that knowing the values, vocabulary, and specific grievances of local people is important in mass recruiting. But since the knowledge to make such specific appeals requires a long association with the people, recruiters usually must rely upon key-men who are native to the area or village to help them tailor their appeals. Through such keymen they attempt to use local customs as well as social and group pressures as tools in winning recruits.

SELECTIVE RECRUITMENT

The recruiting process is dangerous to the underground organization for several reasons. If an individual who is approached informs the police, a valuable recruiter may be lost. Accepting any recruit without investigation and trial may lead to infiltration or the inclusion of undesirables. Recruitment is conducted by a small team which follows the fail-safe principle in each step of the process. A potential recruit is identified and put in a position where he can safely be approached about joining the movement. Then he is led to commit himself through various acts to the underground. He must be tested and trained and finally, if he proves acceptable, he is assigned to a permanent cell.

THE RECRUITERS

At meetings, organizational groups, or discussion clubs, the first recruiting agent, or steerer, identifies individuals who are in positions which might be useful to the underground or who are ideologically susceptible to recruitment. The steerer may engage an individual in general discussion to identify his grievances and feelings toward the government. If an individual is considered ideologically attuned, he is then introduced to the next person in the recruitment chain. The steerer never mentions the underground organization or voices any subversive opinions. It is his job to discover an individual's characteristics and pass the information on. In this way, the steerers conceal their subversive connections and escape the danger of being denounced by recruits whose attitudes they may have misjudged.[1]

THE BUILDUP

The second member of the recruitment team asks the potential recruit to join him at an informal social party or discussion group at which current events and political issues are discussed. In the discussion, the individual's attitudes can be further evaluated and at the same time his background can be checked. The underground agent befriends the recruit and plays upon his ego and personal desires for position, power and importance.

In Malaya, in some of the Communist-led unions, lectures were given to party members several times a week. Agents in the audience observed the reactions of the workers. Those who seemed interested and receptive and who possessed leadership ability were identified as potential recruits. The topics discussed were usually broad social issues rather than ideological material. Front groups were used in a similar way to evaluate the attitudes and inclinations of potential recruits.[2]

COMMITMENT

The individual is not asked pointblank to join but is led gradually toward commitment through a series of small decisions. He is asked whether he is free to distribute leaflets, collect funds, or carry messages. The final decision to join the underground really becomes an extension of lesser decisions preceding it. He may be asked to buy food or other materials from a local store. Later, he is told by a third member of the team that he was buying supplies for the underground and warned that if the police were to find out he would be arrested. He is not given the option of joining, but may merely be asked whether he wishes to stay in the village as a tax collector or leave to join the guerrillas in the mountains. Faced with this choice and implicit threats of violence from the underground or arrest by the police, he will select the least undesirable alternative.

In another approach a person may be asked to donate funds to the movement. If he seems reluctant, the third member of the team may suggest to him that if he collects money from others, he himself need not donate to the movement. The individual may half-heartedly attempt to collect some money and once he has committed himself to this extent, even if he collects only a token amount, he can be coerced into performing other assignments.

THE TEST

The individual's loyalty to the underground is tested by assessing his willingness to perform some minor illegal task. He may be asked to deliver an "important" message to a particular location. The message may be a blank piece of paper with a hidden seal. The individual is then evaluated as to how well he carried out the assignment and whether or not he examined the contents. Even if he reports the incident to the police, nothing is lost.[3] Desire to join is important but often desire wanes; therefore, acts which may be used to coerce loyalty are also required.

THE OATH

Most, if not all, underground movements administer a loyalty oath to new members to impress upon them the seriousness of their jobs and the necessity for secrecy. In most of the European resistance movements, violation of the oath was punishable by death.[4] In the Mau Mau movement, oaths were especially significant, for the Kikuyu tribesmen believed that if they violated the oath they would be punished by a supernatural power.[5]

In the Viet Cong, the ceremony for admission is simple but very serious. A Viet Cong flag and a picture of Ho Chi Minh are used to convey authority. At the swearing-in ceremony the only people present are the applicant himself, the secretary, and two comrades who sponsor the member.

In the Malayan Communist Party, after recruits had been screened in security investigations, they had to be recommended by members with whom they had had contact. Members were often held personally responsible if the recruit proved unreliable.[6]

ASSIGNMENT

During the background check, the recruit's personal record is thoroughly investigated. Until cleared as reliable, he is placed in a probationary cell. During probation, he is tried on various underground activities to determine which he performs best. He is assigned a variety of tasks during this apprenticeship and is forced to practice a variety of security precautions; in this way he is trained to become a well-disciplined, highly security-conscious individual who can be counted on to work independently and to show initiative in future assignments in his permanent cell.

This last step in the recruitment process, assignment to a permanent cell, is reached only after the recruit has been thoroughly tested, observed, trained, and evaluated.

MASS RECRUITMENT

After the underground cadre has been established, a base of support is sought among large segments of the population.[a] Cells set up in the cities and throughout the countryside form the nucleus for action devised to win sympathy and, ultimately, popular support. Most often such support is rallied behind a specific grievance and only later channeled into active insurgency. Many techniques are used to infiltrate mass organizations and gain leadership posts.

FEELINGS OF INDEBTEDNESS

Members of the underground are instructed to create social indebtedness by finding and helping families "in trouble"—the propertyless, the unemployed, and the sick. This enables the underground worker to enter the family or the neighborhood and gain the attention and loyalty of a large number of people. Pressure is then applied to have the people repay their indebtedness by assisting the movement and eventually joining it. A favorable word from a mother, father, relative, or close friend can be a more powerful persuader than any impersonal propaganda message.[8]

The agent, much like a ward or precinct politician, surveys the needs, likes, and dislikes of the people in his district. He may keep individual records on all who live in his area of responsibility. He may find jobs

---

[a] In Indochina, Vo Nguyen Giap said that in order to prepare for an insurrection, organizations must be developed and consolidated within the cities, mines, plantations, and provinces; he stressed that only on the basis of strong political organization could armed organizations be set up.[7]

for the unemployed, arrange housing for those who do not have shelter, or assist farmers with their crops.[b]

In rural areas and small villages, where the close personal contacts among the villagers make it difficult to organize secret cells, a special technique is used. An insurgent force marches into and takes over a village. They assist the farmers in the fields and help raise production, hoping in this way to develop close contact in spite of having come uninvited.[12]

COERCION

Coercion is widely used against those who do not voluntarily join the movement. Techniques range from the simple "armed invitation," where recruiters brandish their weapons and extend an invitation to "volunteers," to more complex techniques of gradual involvement and threat of exposure.[c] Another technique in insurgent-controlled areas is to assign quotas to local officials, such as mayors and village chiefs, who use the social power and prestige of their office to recruit new members.[13]

Even after "recruitment," coercion is continued to discourage defection. Underground recruits are often given money for which a compromising receipt is obtained. They are made to sign documents and papers which would also incriminate them. Even those who are initially attracted by some idealistic approach may lose enthusiasm under stress, but by then they are trapped with evidence of their membership in the movement and have little choice but to remain there.[16]

Mao Tse-tung described what he called the "Road to Yenan" for winning control of the people. He said, "People like doctors, generals, dentists, town mayors, lawyers, who are not rich, do not seek power for itself; much less for the good they can do with it. They want it for the wealth it can bring." He went on to say that if the Communists can help these greedy people, they should. It would be absurd not to help them. The more help they receive the more positions they will help the Communists capture. However, he admonished, never openly participate in

---

[b] The Communist Chinese used mutual aid teams to help farmers harvest crops.[9] Vo Nguyen Giap said, "Our army has always organized days of help for peasants in production work and in the struggle against flood and drought. Political work begins by establishing good relations with the populace."[10] In Cuba, Alberto Bayo recommended that the men should volunteer to repair household items, help put up fences and sow fields, or do any kind of manual work, in order to "demonstrate our affection and gratitude and bring him over to our cause."[11]

[c] A typical example of coercion was used in Malaya. A rubber tapper's duties took him daily into the jungle to tend the rubber trees. On one occasion, he was approached by three armed men. They were friendly, and talked him into bringing them cigarettes the next day. Soon he began smuggling such items as aspirins and flashlights; getting these things past the guards at the village added a little interest to his life. One day, only two of the three men showed up, explaining that the third had been captured and would almost certainly reveal the rubber tapper's smuggling activities. Faced with the alternatives of returning to his village and facing arrest and prison or joining the Communists who would protect him, he joined.[14]

In Cyprus, the EOKA underground used coercive means to secure recruits. A respected watchmaker in Nicosia was warned that his name was high on the EOKA list for execution because he was suspected of having been in contact with the British Army, which he had served with during World War II. He was told, however, that he could save his life and protect his family if he joined the underground. Once in, he became more and more involved. His shop was one of the mail-drops for the underground and he was given increasingly dangerous assignments. He was considered so thoroughly committed to the underground that he was allowed to meet George Grivas, the closely guarded leader of the Greek Cypriot movement.[15]

fraud or plunder, and in carrying out collaboration, never leave evidence that can be used against the Communists. In selecting people, Mao suggests that politicians who have been passed over, doctors mired in mediocrity, and lawyers with limited means be sought out, for, he said, these people know they must fulfill their end of the bargain or be destroyed. The objective is not so much to win friends and sympathizers as it is to gain servants. Workers will stay with you if you get them something and abandon you if you don't.

In countries which have democratic processes and civil liberties, Mao suggests adopting a popular front to attract all groups, leftists, or not, good or bad, sincere or insincere. But above all, "tempt them, each through his particular weakness . . . help them to get what they want, put pressure, first with offers, later with threats. Compromise them if you can, so that they can't get away."

Mao said, "We seek people who serve us, through greed, through fear, inferiority, vengeance, what have you, but who serve us, serve the party, serve the design of the Comintern, serve the cause of the revolution . . . that is the essence of the Yenan way." [17]

SUGGESTION

In recruiting young people, persuasion is particularly effective. Teenagers are highly suggestible and strongly influenced by their peers.[18] It is a usual technique to separate the youngsters from the elders, so that cooler heads will not prevail against youthful enthusiasm.

According to a Viet Cong recruiting document, the first step is to organize large gatherings of young people for a celebration, a political rally, or a cultural event, segregating them from the older people of the village. During the gathering, recruiters make heated speeches denouncing the government. Earlier recruits are planted in the audience to applaud the speeches and volunteer to follow the Viet Cong to the mountain bases. The young people, emotionally aroused by the speeches and seeing fellow teenagers "volunteering," also volunteer to go.

Another technique used by the Viet Cong, in areas where they have little or no following, is to bring in armed propaganda teams. If they have little success in getting volunteers, they kidnap several young men who have not evidenced positive hostility. After training and indoctrination, the young men are returned to their villages where they report on their good treatment. The Viet Cong seldom has difficulty getting subsequent volunteers from the area.

ALIENATION FROM THE GOVERNMENT

Counterinsurgent actions can be used by the rebels to alienate the local people from the government. Insurgent activity may be designed chiefly to draw heavy reprisals, most of which fall on innocent villagers' heads. The insurgents can then point out how unfairly and harshly their government treats its loyal citizens. Recruits will be gained from the vengeful or disillusioned members of the populace.

In Algeria, terroristic action such as assassinations and bomb-throwing caused the French to take severe repressive measures against the general populace. These measures drove many people into the rebel camp.[19] Similarly, in Greece, guerrillas made their attacks on installations appear to come from nearby villages; government troops then retaliated against the villagers. This alienated the villagers from the government forces, and many joined or supported the underground cause.[20]

In South Vietnam, the Viet Cong marched into one hamlet and allowed information to be passed to government security forces that they were going to hold the hamlet for at least 3 days. The next day, Republic of Vietnam military forces arrived and for 18 hours strafed the area from air and ground. When they entered the hamlet, they found that the Viet Cong had long departed; the villagers had borne the losses, both of personnel and property.[21] In the Philippines, one man who was beaten up by government troops as a suspected Huk bitterly joined the Huks in order to get revenge.[22]

Thus, government measures and military actions have alienated a number of individuals and driven them into the insurgent camp. The use of mass destruction weapons such as napalm and artillery on villages believed hiding insurgents is probably sufficient to turn an entire village against the government and toward the insurgents.

APPEALS

Several types of appeals are used to draw people into the movement. Some individuals join because they feel they will receive positions of power as a result of being on the "winning side." An interview with a former anti-Nazi underground member revealed that he joined in hopes of getting a high-ranking post in the government.[23] One undergrounder said:

> . . . The love of power is today endemic even to those social classes which in other days were least susceptible to its temptations. It drives men to run risks and make sacrifices of which they would not otherwise be capable.[24]

One Soviet agent was instructed to search for those who would profit from connections with him. He was instructed not to be concerned with offering material advantages. Instead, he was told to seek out those who were hurt by fate or nature. The types of people to look for are the ugly, those suffering from inferiority complexes, those craving for power and influence, and those who have been defeated by circumstances. He was told to look for people who have suffered from poverty—not so much those who lack material wealth but those who have suffered from the humiliation associated with poverty. Belonging to a strong organization gives them a sense of importance and a feeling of superiority over the handsome and more prosperous people whom they have always envied.[25]

The underground is careful to appear highly selective in its recruiting, thus flattering the prospective recruit with the knowledge that he is wanted by an exclusive group. One former underground member said that he was told that they were interested in him but not his friend; later he found that both he and his friend had been recruited. Another recruit was told that everyone in the movement felt the same as he did, and that if he joined he would have an entire organization at his disposal to help him carry out his own ideas.[26]

Some individuals are enticed into joining the underground in hopes of bettering their condition economically. In Poland during World War II, the underground members never lacked food, clothing, or other necessities, even when such items were generally unavailable.[27] In the Philippines during the Huk insurgency, young men were offered jobs at salaries far higher than could be gotten outside of the movement. Others joined simply because there were no jobs of·any kind available on the outside. In many areas, individuals have prospered by black-market dealings with the underground.[28]

Individual grievances can be amplified to create feelings of helplessness and frustration. The organization is then offered as a means for redress. The individual is led to believe that by joining he will receive the massive support of a large organization to remedy his personal grievance.

It is clear, therefore, that effective appeals for recruiting underground members are many and varied. One individual may feel that the cause is good and join the movement out of conviction; such a decision is usually based upon careful deliberation. A second individual may join as an emotional reaction against people or conditions: he identifies himself with the leaders or members of the movement and once committed may use its goals and ideas as a substitute religion. A third type of individual is influenced by social pressures and joins the movement because his parents, friends, relatives, or neighbors are members. A fourth type seeks personal advantage. Another may join because of a "bandwagon" effect; if the movement is succeeding, he may join because others join or because he fears being penalized or persecuted in some way if he does not join before victory is obtained.[29]

DEVELOPMENT OF KEYMEN

In areas where the agent must remain secret or where the insurgent underground has no control, the agent-organizer is instructed to form secret channels consisting of himself, a keyman, and sympathizers. He may recruit up to three keymen, each of whom recruits two or three sympathizers, who in turn contact the people of the village.

The type of keyman sought by the agent is one with natural leadership abilities, the respect of his community, and some susceptibility to recruitment. The approach to this individual is tailored to his dissatisfactions. The most important thing that the agent can do is to be a sympathetic listener, hearing complaints and using these details to build up a sense of dissatisfaction. According to one captured member of the Viet Cong:

You must be patient in listening to people's problems. You must know what they are talking about and, most of all, you must leave the impression that their specific grievance is your main concern. With the peasants you discuss land reform and perhaps you promise education for the youth. You do not go far on generalities.[30]

Thus, the first step of the agent in his recruitment of keymen is to survey the conditions and attitudes of the group from which he seeks recruits. Once a natural leader has been spotted, the second step is to involve this potential keyman in some campaign, preferably one related in some way to his personal grievance. Once his interest has been aroused, he is given minor assignments. The third step is to change the individual's specific dissatisfaction into general unrest regarding the status quo. All current evils are blamed on the government. Where there is overlapping loyalty to the government and to the movement, the agent must polarize it, build up an "in-group" feeling, and alienate the recruit from the government. In doing this, the ruling authority is always referred to abstractly; government leaders are not attacked personally.

Once he is recruited, the keyman in turn recruits sympathizers and is asked to infiltrate and seek leadership positions in civic organizations. His recruiting is clandestine until a cohesive group has been developed. Then the collective support of the organized group serves to encourage others; there is a bandwagon effect—other members in the community join because their friends are members or because of the social pressure brought to bear by the group.

Keymen are important because people are more willing to accept guidance from members of their own community than from outsiders. Further, a keyman knows the special conditions within his community and can phrase appeals within its context. Since there is more confidence in the keyman than in the agent-organizer, it is the keyman who communicates with the people, not the organizer.

In summary, underground recruitment techniques are probably most successful when selectively applied. To a large degree, underground recruitment depends upon the careful screening and constant testing of potential recruits. This is true for both selective and mass recruitment. Careful study of potential recruits' loyalty to the government and their personal likes and dislikes is important in all underground recruitment.

FOOTNOTES

[1] Lucian W. Pye, *Guerrilla Communism in Malaya* (Princeton, N.J.: Princeton University Press, 1956), pp. 220–21.

[2] *Ibid.*, pp. 173–75 and 218ff.

[3] Interview with a former member of the Polish underground.

[4] Andrew R. Molnar, *et al.*, *Undergrounds in Insurgent, Revolutionary and Resistance Warfare* (Washington, D.C.: Special Operations Research Office, 1963), p. 79.

[5] Frank Kitson, *Gangs and Counter-Gangs* (London: Barrie and Rockliff, 1960), p. 128.

[6] Pye, *Guerrilla Communism*, pp. 243–44.

[7] Vo Nguyen Giap, *People's War, People's Army* (Washington, D.C.: Government Printing Office, 1962), pp. 77–78.

[8] Fred H. Barton, *North Korean Propaganda to South Koreans (Civilians and Military)*, Technical Memorandum ORO–T–10 (EUSAK) (Chevy Chase, Md.: Operations Research Office, February, 1951), p. 110.

[9] Floyd L. Singer, *Control of the Population in China and Vietnam: The Pao Chia System Past and Present* (China Lake, Calif.: U.S. Naval Ordnance Test Station, November, 1964), p. 9.

[10] Giap, *People's War, People's Army*, pp. 55–56.

[11] Alberto Bayo, *150 Questions for a Guerrilla* (Boulder, Colo.: Panther Publications, 1963), p. 53.

[12] *Ibid.*, p. 85; and Giap, *People's War, People's Army*, p. 56.

[13] Alvin H. Scaff, *The Philippine Answer to Communism* (Stanford, Calif.: Stanford University Press, 1955), p. 121; and William J. Pomeroy, *The Forest: A Personal Record of the Huk Guerrilla Struggle in the Philippines* (New York: International Publishers, 1963), p. 43.

[14] Richard L. Clutterbuck, "Why Chi Keong Surrendered," *Marine Corps Gazette*, XLVIII (July 1964), pp. 32–36.

[15] Dudley Barker, *Grivas: Portrait of a Terrorist* (New York: Harcourt, Brace, and Co., 1959), pp. 140–41.

[16] J. Edgar Hoover, *Masters of Deceit* (New York: Henry Holt, 1958), p. 300.

[17] Eudocio Ravines, *The Yenan Way* (New York: Charles Scribner's Sons, 1951), pp. 151–57.

[18] Robert R. Blake and Jane S. Mouton, "The Experimental Investigation of Interpersonal Influence," *The Manipulation of Human Behavior*, eds. Albert D. Bidderman and Herbert Zimmer (New York: John Wiley and Sons, 1961), p. 251.

[19] Paul A. Jureidini, *et al.*, *Casebook on Insurgency and Revolutionary Warfare: 23 Summary Accounts* (Washington, D.C.: Special Operations Research Office, 1962), p. 255.

[20] Interview with a former member of the Greek EAM.

[21] Edward G. Lansdale, "Vietnam: Do We Understand Revolution?," *Foreign Affairs*, XLIII, No. 1 (October 1964), p. 85.

[22] Scaff, *Philippine Answer*, p. 119.

[23] Interview with a former member of the anti-Nazi German Social Democratic group, the "New Beginning."

[24] A. Rossi, *A Communist Party in Action* (New Haven, Conn.: Yale University Press, 1949), p. 216.

[25] Nikolai Khokhlov, *In the Name of Conscience* (New York: David McKay, 1959), p. 164.

[26] Interview with a former member of the anti-Nazi German Social Democratic group, the "New Beginning."

[27] Jan Karski, *The Story of a Secret State* (London: Hodder and Stoughton, 1945), p. 63.

[28] Scaff, *Philippine Answer*, pp. 108–109.

[29] Rudolf, Heberle, *Social Movements* (New York: Appleton, Century, and Crofts, 1951), pp. 95–100.

[30] *The Washington Post*, May 19, 1964.

CHAPTER 7

EDUCATION AND TRAINING

Educational and training programs are an essential organizational feature of underground movements. The importance of training special cadres for the successful launching and operation of insurgencies has been emphasized in most undergrounds.

SCHOOLS

The training activities of the international Communist movement are particularly illustrative of underground training. The Communists have long specialized in the establishment of special schools for providing "international instructors." These schools have frequently been established on an international basis, outside of the countries involved. During the early phase of the Comintern, the primary training headquarters for the Communist movement was the Soviet Union. The University of the Workers of the East, located near Moscow, was established in 1921 to train revolutionaries selected from throughout the world for special instruction. The school trained agents in underground political tactics, guerrilla warfare, intelligence and organization methods, and the promoting of agitation and strikes. Students at the school used cover names and addresses, so that when they returned to their own countries they could instruct others in subversive methods without danger of exposure.[1]

Today, the Lenin School near Moscow provides advanced training to Communists who have proven themselves in national parties. Again, the students follow the rules of conspiratorial behavior in their day-to-day activities, with assumed names and false biographies. The courses vary in length from 1 to 3 years, and the subjects covered include ideology, mass agitation, strikes, and guerrilla tactics. During the summer the classes move to the field for special exercises, mapreading, and weapons training. Prominent Communists, such as Stalin and Tito, have lectured at the school. One well-known graduate of the Lenin School is Walter Ulbricht of East Germany.[2]

There are other schools where young students receive political and ideological training. The Communist University for Western National Minorities (KUNMZ) was established in Moscow in 1921 to train White Russians and Ukrainians living outside the Soviet Union. A German sector was created after 1933 when large numbers of German Communists emigrated to the Soviet Union. Over a 3-year period, the youths covered Marxism-Leninism, party history, dialectical materialism and other political and cultural subjects. One year before the school closed, there were 250 students enrolled, approximately half of whom were German or Austrian.[3]

The People's Friendship University in Moscow was established to "train leaders for new countries of Africa and the poorer, older ones of Latin America." Its colleges concentrate on psychological-political warfare. One former Communist trainee reports that the range of instruction in political warfare subjects includes:

General

The doctrines of Marx and Lenin concerning the role of government (the state, the party, and their roles in society);
The Communist Party insurrectionary organization, its structure and methods;
Labor unions as an instrument of economic and political warfare against capitalist democratic society;
The strategy of neutralizing or demoralizing the middle classes;
The strategy of winning over or neutralizing the farm population of the advanced countries;
Communist colonial policy—the teachings of Lenin as elaborated by Stalin;
The peasants as a main base for igniting colonial revolution.

Underground Warfare

The role of Communists in the event of war against the Soviet Union;
Infiltration of armed services;
The relation between aboveground (legal) and underground (illegal) activities and the necessity of carrying on both at the same time;
The purpose and methods of infiltrating government departments;
The role of sabotage and espionage in political warfare.

Armed Insurrection

How to form a paramilitary combat force;
Means and methods of arming such a force;
The role of such a force in case of war against the Soviet Union;
The general scheme of seizing a city;
How to hold a city after seizure;
The supremacy of surprise in carrying out a successful insurrection;
Techniques and objectives of guerrilla warfare;
Probable countermeasures of a government sensing an insurrection and methods of overcoming same;
The consolidation of power.[4]

It has been reported that several thousand men and women have been trained in these colleges and then assigned to various posts throughout the world. Those from foreign areas are assigned to the countries from which they came, but can be recalled to Moscow at any time the university desires.[5]

As interest grows in new areas, new subjects are introduced and tailored to the situation. For example, the potential use of witchcraft is included in programs for African nations. Reportedly, native Africans are trained in witchcraft so that upon their return to Africa they can use it to create unrest and cause the population to rise against the Caucasians.[6] One student who had participated in such a course but who later defected, stated that he was instructed in how to produce speech or noises from a skull or a skeleton by the use of hidden microphones, how to simulate spirit rappings, how to make a phantom appear from a cloud of smoke, and other bizarre techniques.[7]

In reviewing the activities of past and present Moscow schools and the Communist undergrounds organized since the 1930's, there is evidence that many underground leaders have been trained in the Soviet Union. Besides Marshal Tito of Yugoslavia and Walter Ulbricht of Germany, these include Maurice Thorez of France, Crisanto Evangelista of the Philippines, Lai Teck of Malaya, Ho Chi Minh of Indochina, Chou En-lai of China, and Tan Malakka, Muso, and Alimin of Indonesia.[8] Muso remained more than 20 years in the Soviet Union. He was said to have been Stalin's choice as chief Communist leader in Southeast Asia. Both Muso and Alimin attended Lenin University, where the latter met Earl Browder of the United States, Pollitt of Great Britain, Sharkey of Australia, Thorez, and Chou En-lai.[9] There were, of course, many others such as Ana Pauker of Rumania who, because of their training in the Soviet Union, were chosen for positions of importance in post-World War II Communist governments.

The Moscow schools and universities were significant because they inculcated students with the importance of the policies of the Communist Party. In addition, they provided the technical knowledge necessary to run secret organizations throughout the world and to bind them together, both on a personal and organizational level. The graduates also served as a backup for Russian intelligence, one of the principal reasons for establishing schools for foreign students in the USSR.[10]

The idea of training special cadres has spread from Moscow and schools have been established in many other parts of the world. For example, the Chinese Communist Party maintains a number of schools for pupils from Latin America. In a conference between Mao Tse-tung and Khrushchev in Peking it was agreed that the Chinese had more experience than the Russians in guerrilla warfare and therefore could more effectively operate training centers for South American needs.[11]

Alberto Bayo, Castro's teacher and mentor, directs a training site in Cuba which acts as a center for Latin American subversion. Bayo learned his lessons in tactics as an officer in the Spanish Foreign Legion. While serving in Morocco he became impressed with the ability of a few guerrillas to harass columns of troops many times their number. Bayo made a study of guerrilla warfare and advocated that it be used by the Spanish Republican forces (with whom he later served) against the forces of General Franco. His superiors demurred, however, because "conventional warfare" was the order of the day. With the defeat of the Republicans, Bayo fled to Cuba, and later to Mexico, where he trained exiles from Nicaragua and the Dominican Republic who made abortive attempts to start revolutions in their respective countries.[12]

After his initial failure in Cuba in 1953, Fidel Castro went to Mexico and prevailed upon Bayo to train his first band of 58 men. This group underwent a rigorous 6-month course which included long daily marches with full pack, traversing jungle country at night, firearms practice, and the manufacture and use of all types of demolition equipment.

Alberto Bayo now heads nine training schools for revolutionaries in Cuba. A Venezuelan student in one of these schools has stated that during the 4-month course the students worked 16 hours a day, 7 days a week. The principal textbook was Bayo's book, *150 Questions for a Guerrilla*. The students were taught to make various kinds of incendiary and time bombs, boobytraps, mines, and torpedoes. They also learned specific techniques for destroying bridges, oil pipelines, communication facilities, police stations, and even large government buildings.[13]

Sabotage, however, was only part of the course. Emphasis was also placed on terror tactics to be used in urban areas to provoke riots and incite mobs. Techniques of robbing banks, destruction of natural resources, and assassination were also on the curriculum.[14]

On July 3, 1963, the Council of the Organization of American States called attention to the existence in Cuba of a training center for subversive activities sponsored by international communism. The instructors for this center came from the USSR, China, and Czechoslovakia, while most of the students were from Latin American countries. The Council foresaw not only the danger of the ideology taught at the Cuban center, but the even greater danger in the eventual return of students to their own countries as underground agents.[15]

External training bases are common to most insurgent movements. The Algerian Front for National Liberation (FLN) followed the Communist pattern. Headquartered in Cairo, the FLN early made arrangements for establishing training bases for Algerian revolutionaries in Libya, Tunisia, and Morocco. Training was also provided in Egypt, Syria, and Iraq.[16]

In Greece, an external training base for Communist guerrillas was located in Yugoslavia. A report issued in 1947 by the United Nations Special Commission on the Balkans mentioned the existence of a training camp at Bulkes, Yugoslavia.[17]

The anti-Portuguese insurrection in Angola began in March 1961. Angolese insurgents were sent to Tunisia where they undertook a 7-month training course with the Algerian Liberation Army. Upon their return, 18 were appointed officers and given the task of training the entire rebel army. The Congolese provided an abandoned bivouac area about 70 miles from the Angolan border for the site of the training base. By November 1963 the camp was reportedly training 2,200 men every 8 weeks. In addition to weapons training, emphasis has been placed on political indoctrination.[18]

External sources also influenced the training of the Vietminh in the early 1950's. Mao Tse-tung's scheme for conducting victorious wars of national liberation was outlined at a conference in Peking in November 1949, attended by Vietminh members and Communists from other Southeast Asian countries.

The Vietminh quickly adopted the Mao formula and General Giap returned to North Vietnam in the 1950's with two basic plans. The first ordered general mobilization while the second made provisions for a change from purely guerrilla warfare to mobile warfare. While in China, General Giap also made arrangements for Vietminh guerrillas to be trained as regular forces in Kwantung. Accordingly, early in 1950, thousands of Vietminh traveled to China. As the Mao influence increased, Ho Chi Minh introduced far-reaching political changes. To emphasize the need for a long-term struggle, he personally translated Mao Tse-tung's *On Protracted Warfare* into Vietnamese.[19]

Much of the present-day training of the Viet Cong takes place in North Vietnam. One of the principal training centers is Xuan Mai near Hanoi. Political and military subjects are taught and training is given in such specialties as metallurgy, medical treatment, and intelligence work. The courses range from 4 to 6 months. The trainees then move to Vinh on the east coast and stop for additional training at Don Hoi, from where they are trucked to the Laotian border. From Laos they infiltrate South Vietnam and implement their training.[20]

TRAINING

LITERATURE

An international body of literature on the strategy and tactics of modern insurgency, underground, and guerrilla warfare has markedly increased during the past half century. Communist writers have perhaps contributed most to the literature. One of the first was Lenin, who in 1901 began writing "What Is To Be Done?" followed later by *State and Revolution* and *Left-Wing Communism*.[21] These works are not remote philosophical essays, but concise statements of the strategy and tactics of revolution. Lenin called for an integration of organizational and combat factors with Marxist economics and sociology, and set forth a guide to such necessary organizational and political work. Similarly, Mao Tsetung's writings [22] serve as a primer for revolutionary strategy and tactics, emphasizing the political as well as the role of purely guerrilla-type forces in insurgent action. Both Lenin and Mao Tse-tung stress the close relationship between political and military or guerrilla action. Their works have been reprinted in millions of copies and are used as basic training manuals in many countries.

Following Mao Tse-tung's lead, General Giap of North Vietnam elaborated on the techniques and strategy of guerrilla warfare in *People's War, People's Army*. General Giap stresses the coordination of political, propaganda, and military strategies during insurgencies. He calls this "armed propaganda," advising that "political activities [are] more important than military activities, and fighting less important than propaganda." General Giap's lists of principles and advice for "preparing forces for an insurrection" make it an important modern handbook for insurgents.[23]

Similarly, Ernesto "Che" Guevara in *Guerrilla Warfare* relates in detail various procedures and tactics developed during the Cuban insurrection of 1959. Although influenced greatly by the works of Mao Tse-tung, Guevara tailors his strategies to fit the Latin American environment. He deals more with daily tactics than with overall organization or planning.[24]

More pragmatic still is Gen. Alberto Bayo's *150 Questions for a Guerrilla*. In terse, specific fashion, this experienced insurgent spells out the most mundane elements of organizing and executing underground and guerrilla warfare. For example, he explains how a camp is best organized and manned, down to latrine orderlies. Bayo also prepared a kind of do-it-yourself kit for insurgents, explaining multifarious techniques and giving detailed instruction for making such things as "Molotov cocktails," tank traps, mines, and boobytraps.[25]

An example of a non-Communist handbook for guerrillas is Abdul Haris Nasution's *Fundamentals of Guerrilla Warfare*.[26] This Indonesian army officer reports on the experiences of Indonesia's insurgency against the Dutch. He lists a number of "fundamentals" for insurgent operations, enumerating the kinds of political, psychological, and military factors that should be considered.

Because these writers, particularly Giap, Guevara, and Bayo, emphasize action as well as ideological analysis, their works have been used as training manuals for both anti- and pro-Communist forces.

POLITICAL INDOCTRINATION

Political and ideological training in Communist movements has importance over and beyond its content. Such training imbues the individual with a sense of dedication and ideological purpose that will insure his carrying out all directives, even under conditions where the party has no control.[27] The more distant a unit is from central control, the greater is the political training.

In the Viet Cong military organization, for example, full-time regular units have a rigid training schedule in which two-thirds of the time is spent in military study and one-third devoted to political content. At the district level, the proportion is fifty-fifty for political and military study. At the village level, study is 70 percent political and 30 percent military. The units are required to study the scope and objectives of the National Liberation Front (NFLSV) as well as the guerrilla warfare tactics of Mao Tse-tung.[28] Political training includes discussions of communism, plans for winning control of the country, and the need for support of the National Liberation Front. They are instructed to study each issue and are then required to "adopt an attitude." This is designed to enable each man to react immediately in a politically "correct" manner on every question.

The aim of Communist political instruction is to imbue students with the proper attitudes, beliefs, and objectives of the Communist system. Thus trainees learn to apply the "correct" approach even in the absence of directives.

ORGANIZATION

One of the most important aspects of Communist training is its emphasis on organizational techniques. A former Communist who was in charge of the Communist Party's Latin American bureau has described the party's training approach. His first task upon arrival in Bogotá, Colombia, was to set up courses in organizational work for local labor unions and peasant groups. Organization was the aspect Latin American leaders knew least about, and they were greatly impressed with anyone who could bring them organizational knowledge. He trained ten people at a time during a series of 2-week courses in organization, and by the end of the year his trainees' organizations had all increased their membership. Later, he established courses for training students in organizational techniques at a university in Caracas, Venezuela.[29]

An American and former Communist who served with the Huks' National Education Department in the Philippines described the daily activities of this center of revolution. One of the important features of the department's work was to publish, twice a month, 4- to 6-page self-study booklets, one for Huk soldiers and one for political workers, covering such subjects as organizing people and operating schools.[30]

GUERRILLA TRAINING

The study of guerrilla warfare principles, particularly Mao Tse-tung's protracted war thesis, is also stressed in many Communist training programs. Communist doctrine suggests that military training should be carried out every day and that new techniques should be introduced regularly. Frequently, trainees are instructed to discuss and draw conclusions from field exercises and from documents exchanged between regions, and they are admonished to carry out battle analysis after every engagement.

The Vietminh training program varied for the different levels of military forces. The local or village level units had a self-training program consisting of political, small arms, and sabotage instructions. More advanced village units performed close-order drill and received automatic weapons instruction. Assistance for these training programs was provided by regional or occasionally main-force troops.[31] At the district level, additional instruction on the use of automatic weapons and individual arms was given and fundamentals of small-unit tactics were introduced.

The regional units usually were made up of men who had been in the local and district guerrilla units and had received basic training with those units. At the regional level, more emphasis was placed on individual instruction on advanced weapons, unit tactics, special skills, and low-level staff duties. Instructors were usually officers from the main force. The men of the main force were usually chosen from the regional guerrilla units.

The process of training and rising through the levels of organization often took several years and political training was stressed throughout.

The main objective of the training program was to produce politically reliable and enthusiastic soldiers who could serve both as experienced fighters and propaganda agents.

UNDERGROUND SUPPORT TRAINING

North Vietnamese have trained infiltrators to manufacture and repair crude weapons to make simple blast furnaces for producing cast iron for weapons. Specialists, such as doctors, pharmacists, union and youth organizers, and radio technicians have also been given 2 months of basic training before infiltrating south.[32]

Training for the Viet Cong has included special intelligence instruction in radio transmission, coding and decoding, use of ambush techniques, sabotage tactics, methods for enlisting draft-evaders, and terrorist techniques.

Training in propaganda techniques is also stressed. One ex-Communist has described how trainees learned to produce propaganda leaflets under clandestine conditions. The methods and ways of manufacturing materials needed for production were explained. They were instructed in methods of production, such as using a lump of clay to produce up to 100 copies of short text. They were also shown photographic methods of reproducing leaflets and newspapers, including drawings and caricature.

Although the instruction was very detailed, trainees were not permitted to take notes and were required to memorize everything. Reportedly, this was to give the students practice in clandestine behavior.[33]

Courier activities are also frequently stressed in training for underground operations. For example, the non-Communist underground movement in Czechoslovakia in 1952 operated a special school in the Tatra Mountains in Eastern Slovakia to train couriers. Trainees in this program were required to run until they felt exhausted, and then run another mile. They were taught to swim rivers with their clothes on in zero temperatures and, having crossed, to run in order to keep their clothes from freezing to their bodies. They were also taught to go into hiding for a week without food, in both summer and winter. Emphasis was placed on the identification of contacts. They were, of course, warned against gambling, heavy drinking, and women, since the Communist regime utilized women as decoys.

Once the students were considered thoroughly trained, they were given easy missions, such as making contact with other underground members; finally they were sent on more dangerous missions in which they had to cross the Czech border into Austria and return, bringing with them materials needed by the underground. The couriers had to avoid fishermen on both the Danube and Morava Rivers because the Communists issued fishing permits only to those who took an oath to turn in border-crossers, and they had to avoid well-traveled trails on the Austrian side as the Czech authorities offered large rewards in Austrian currency to anyone who caught a border-crosser on Austrian soil.[34]

Besides political indoctrination of cadres, "trial by fire" seems to be an important aspect of all underground training. Cadres must be tried and tested in the field. According to Communist theory, work in the field not only increases an agent's knowledge, but gives him an opportunity to exercise leadership in a specific situation. Further, field activities deepen commitment and test reliability and capabilities.[35] While ideology may be the chief factor for joining, Communists do not believe that ideology alone will sustain commitment to the organization. Hence, constant participation in activities is emphasized.

In summary, education and training are essential elements in underground administrative operations. Undergrounds need schools in order to train cadre in the tactics, techniques, and strategy of underground operations and methods; most frequently these schools are located in a foreign nation outside of the area of conflict. Underground training particularly emphasizes organizational skills and political training, equipping its cadre to make independent, rapid decisions in the field. Indeed, to make underground decentralization operative, such training is essential. Finally, underground training provides the skills and knowledge necessary for members in adapting to the requirements of clandestine work.

FOOTNOTES

[1] Gunther Nollau, *International Communism and World Revolution* (New York: Praeger, 1961), pp. 165–67 and 171–77.

[2] *Ibid.*, pp. 172–73.

[3] *Ibid.*, pp. 174–75.

[4] William R. Kintner and Joseph Z. Kornfeder, *The New Frontier of War* (Chicago: Henry Regnery, 1962), p. 36.

[5] *Ibid.*, p. 37.

[6] Willard Edwards, "African Witchcraft Training Described," *The Washington Post*, November 28, 1964.

[7] *Ibid.*

[8] Andrew R. Molnar, *et al.*, *Undergrounds in Insurgent, Revolutionary and Resistance Warfare* (Washington, D.C.: Special Operations Research Office, 1963), p. 153; and Anthony T. Bouscaren, *Imperial Communism* (Washington, D.C.: Public Affairs Press, 1953), pp. 68–71.

[9] *Ibid.*, p. 71.

[10] Nollau, *International Communism*, p. 176.

[11] U.S. Congress, Senate, Committee on the Judiciary, Subcommittee to Investigate the Administration of the Internal Security Act, *Hearings on the Communist Threat to the United States Through the Caribbean*, 86th Cong., 1st Sess., 1959, pp. 55 and 68.

[12] Alberto Bayo, *150 Questions for a Guerrilla* (Boulder, Colo.: Panther Publications, 1963), pp. 8–9.

[13] Juan DeDios Marin, "Inside a Castro 'Terror School'," *The Reader's Digest* (December 1964), pp. 119–20.

[14] *Ibid.*

[15] Council of the Organization of American States, Special Committee, *Report* (Washington, D.C.: Pan American Union, July 3, 1963), pp. 10–11.

[16] Paul A. Jureidini, *et al.*, *Casebook on Insurgency and Revolutionary Warfare: 23 Summary Accounts* (Washington, D.C.: Special Operations Research Office, 1962), p. 253.

[17] Harry Howard, *U.N. and Problems of Greece*, Publication 2909 (Washington, D.C.: Department of State, 1947), p. 16.

[18] Lloyd Garrison, "Now Angola: Study of a Rebel," *The New York Times Magazine* (February 16, 1964), pp. 12–13ff.

[19] Denis Warner, *The Last Confucian* (New York: Macmillan, 1963), p. 40.

[20] U.S., Department of State, *Aggression from the North: The Record of North Vietnam's Campaign to Conquer South Vietnam* (Washington, D.C.: Department of State, 1965), p. 5.

[21] Vladimir I. Lenin, *Collected Works* (Moscow: Foreign Languages Publishing House, 1961).

[22] Mao Tse-tung, *Selected Works*, (London: Lawrence and Wishart, 1954).

[23] Vo Nguyen Giap, *People's War, People's Army* (Washington, D.C.: Government Printing Office, 1962).

[24] Ernesto Guevara, *Guerrilla Warfare* (New York: Monthly Review Press, 1961).

[25] Bayo, *150 Questions*.

[26] Abdul Haris Nasution, *Fundamentals of Guerrilla Warfare* (New York: Praeger, 1965).

[27] Philip Selznick, *The Organizational Weapon: A Study of Bolshevik Strategy and Tactics* (Glencoe, Ill.: The Free Press, 1960), p. 19.

[28] U.S., Department of State, *A Threat to the Peace: North Vietnam's Effort to Conquer South Vietnam*, Part I (Washington, D.C.: Department of State, 1961), pp. 8, 101, and 102.

[29] U.S. Congress, Senate, *Communist Threat to the United States*, p. 44.

[30] William J. Pomeroy, *The Forest: A Personal Record of the Huk Guerrilla Struggle in the Philippines* (New York: International Publishers, 1963), p. 146.

[31] George K. Tanham, *Communist Revolutionary Warfare* (New York: Praeger, 1961), pp. 59–64.

[32] U.S., Department of State, *Aggression from the North*, pp. 8–10.

[33] Wolfgang Leonhard, *Child of the Revolution* (London: Regnery, 1957), p. 188.

[34] Stefan Ilok and Lester Tanzer, *Brotherhood of Silence* (Washington, D.C.: Robert B. Luce, 1963), pp. 5, 8.

[35] Selznick, *Organizational Weapon*, p. 19.

CHAPTER 8

FINANCE

An underground organization needs funds to carry out its activities. Full-time agents must be paid and military units armed; escape and evasion networks must have money for extra food, for safe houses, and to give to escapees; psychological operations need funds for publications, visual aids, and portable radios; headquarters and administrative sections need typewriters, radios, and so on. The exact income needs of an underground naturally depend on the nature and scope of its operations.

COLLECTION METHODS

When a government-in-exile is associated with the underground movement it can provide a symbol of legalism; it can negotiate substantial loans from other governments, issue bonds, establish currency, and perform similar fund-raising tasks.

Collection methods in underground movements vary with the source being tapped. Funds from sources outside of the country—foreign governments, expatriates, foreign sympathizers, business speculators—are usually solicited by small teams of collectors. The funds collected are transferred back into the country through couriers, international banks, or dummy corporations set up by the underground.

During the Algerian revolution, the FLN sent fund-collecting teams to Arab and European countries. In Arab towns the imam or other religious leader was contacted and requested to plead for the rebel cause or allow the team members to do so. Using the religious setting to advantage, emotional appeals for Arab brotherhood were made. The congregation was then asked to contribute to the FLN. The imam was given a percentage of the collection—sometimes as much as half—in return for sanctioning the collection. Implicit threats of retaliation were used to deter the collectors from taking funds for themselves, but as long as the net amount was satisfactory, the leaders ignored slight discrepancies. The funds were taken to Tangier and eventually deposited in numbered accounts in Spanish or Swiss banks.[1]

Often the underground movement establishes a central finance collection agency to acquire funds from sources within the country. In Malaya, for example, the Min Yuen had this responsibility; they extorted cash from landowners, mine operators, and transport companies, and "taxed" workers in the local plantations and tin mines.[2]

In the Philippines, the Communist underground apparatus organized its finance department on three levels—national, regional, and district. Each district finance department had an accounting section with an accountant, bookkeeper, and cashier, and a "contactmen and collectors" division.[3]

Foreign governments often assist if an underground movement is in opposition to a common enemy. Sometimes a foreign government contributes so that if the movement is successful it can expect some reciprocity from the new government.

Wealthy individuals or commercial enterprises may contribute voluntarily to the movement. For instance, the Hukbalahap movement in the Philippines and the Malayan Min Yuen received substantial contributions from wealthy businessmen in Manila and Singapore.[4]

The underground can solicit for loans among the population and business community, but it must deal with the problem of establishing the authenticity of the collecting agent and the reliability of the organization itself. In most cases, some form of IOU is offered.

Private friendship societies or quasi-official aid groups often assist an underground when their special interests are involved. The international labor movement supported the early anti-Nazi underground. One of the more celebrated cases of support was the Jewish Agency's support of the Palestine revolution. Throughout the western world, the agency established officers or representatives to make open appeals for money in newspapers, at lectures, and at social events.[5]

An underground may raise money by selling various items, including narcotics and fraudulent lottery tickets. One underground unit in the Philippines raised money by having dances and charging admission. The Min Yuen sold stolen rubber and tin on the black market and is said to have collected a million and a half British pounds for its efforts. The underground may conduct sales either door-to-door or through "front" stores. The Malayan Communist Party operated a bookstore, coffee shops, and general stores.[6] Undergrounds sometimes resort to such measures as bank and payroll robbery and train hijacking. Some have special units for this purpose: The Blood and Steel Corps of the Malayan Communist Party was an example of this.[7]

A system of taxation can provide substantial funds for an insurgency. The Viet Cong tax system calls for an economy-finance committee to be established in each provincial capital, and a specially selected collection committee in each town, village, and hamlet. The collection committee's activities are divided into three phases: investigation, consultation, and collection. First, the committee investigates in detail the occupation and annual income of each individual in its assigned area. The individual's normal annual production, in appropriate units, is ascertained and divided by the number of members in the family, including any members who are presently with the Viet Cong forces or who have been killed or imprisoned by the government forces. For instance, if a family consists of father, mother, one son at home, and two sons with the Viet Cong, and if its total annual production amounts to 60 units of grain, the total is divided by five. This quotient is then assessed in money at the going rate and taxed according to an established percentage. The greater the annual production, the greater the tax.

After the estimated tax has been figured for each householder in the village, the collection committee begins the "consultation" phase by holding meetings and lecturing the villagers. Agents planted in the audience support the lecturers and spontaneously volunteer to pay in the name of patriotism. If an individual does not volunteer, he is taken aside and privately asked how much he is willing to donate. If he decides to donate an amount higher than the committee's assessment, it is accepted. If the amount is lower than the committee expected to collect, tax information is used to intimidate him into a higher pledge. Throughout the discussions, the implicit use of force is present.

This system has apparently been effective in the areas controlled by the Viet Cong, but considerable difficulty is experienced in areas controlled by government forces, where the people have already been taxed by the government. To overcome this, propaganda is intensified to convince the people that the Viet Cong will eventually win the war and that they should support the future, rather than the present, government.[8]

In some cases a subversive movement can control transportation routes and collect tolls. The Viet Cong operate toll booths on arterial Route 1, 50 miles east of Saigon. Nearly all cars, buses, and trucks are stopped and charged $2.00 to $10.00, depending on the weight of the vehicle and the cargo. In a 4-month period from November 1964 to February 1965, an estimated $40,000 was collected.[9]

The use of local currency facilitates exchange for local goods and services, but the physical transfer of the money presents a problem, and the government may take countermeasures by replacing the local currency with scrip.[10] Substitutes for local currency range from IOU's to U.S. dollars or British pounds; in the Congo, travelers' checks were used in lieu of local currency. The Viet Cong use North Vietnamese currency in some areas in South Vietnam and issue their own money in other areas. The underground civil government in the Philippines issued its own scrip in some instances.[11] It has been said that it is possible to determine how the movement is progressing on the basis of commercial investment and the use of rebel currency.

HUMAN FACTORS CONSIDERATIONS

People may contribute to an underground for a variety of reasons: allegiance to the cause, social pressures, present or future self-protection, chance of personal gain, or a desire to be on the winning side.

Some people who have been influenced by the movement manifest their support through regular and voluntary donations. Individuals confronted with social pressures to contribute to a movement may find it difficult to refuse to comply. Applied social pressure is seen in the FLN's use of Muslim religious leaders to make pleas for money within the Muslim community and in the Viet Cong tax-collection system of planting enthusiastic "volunteers" in an audience to pressure others into contributing.

Businessmen may contribute to a movement as an investment in or a hedge against the outcome of the revolution, so that if the movement is victorious they can be identified and treated as supporters. Some people find it profitable to deal with an underground. An underground may willingly pay inflated prices for various items obtainable only through the black market. Individuals and firms sometimes make loans and contributions to an underground with stipulations for later concessions from the movement if and when it gains control of the government. A foreign firm allegedly gave weapons to the Algerian underground in return for a pledge that it would be given the oil concession in the country when the revolutionaries took over the government.[12]

If the underground employs coercive means to collect funds, individuals and business concerns may contribute for their own protection, paying the minimum acceptable amount to avoid reprisals. People who have once contributed, for any reason, may continue to contribute upon threat that their initial support will be revealed to the government.

People tend to be more amenable if there is an indication of return on their investment. Even a simple IOU helps, and underground members are usually directed to give some form of IOU. There are less strenuous objections to taxes imposed by an underground if they are levied with apparent fairness, as in the Viet Cong's impartial production unit assessment system.

FOOTNOTES

[1] Interview with a former member of the Algerian FLN collection group.

[2] Andrew Molnar, et al., *Undergrounds in Insurgent, Revolutionary and Resistance Warfare* (Washington, D.C.: Special Operations Research Office, 1963), pp. 253–54.

[3] Fred H. Barton, *Salient Operational Aspects of Paramilitary Warfare in Three Asian Areas*, ORO–T–228 (Chevy Chase, Md.: Operations Research Office, 1953), p. 162.

[4] Molnar, *Undergrounds*, pp. 254 and 322.

[5] Interview with a former member of the anti-Nazi German Social Democratic group, the "New Beginning"; Molnar, *Undergrounds*, p. 340.

[6] Lucian W. Pye, *Guerrilla Communism in Malaya* (Princeton, N.J.: Princeton University Press, 1956), p. 80.

[7] *Ibid.*, p. 88.

[8] Peter Grose, "Vietcong's 'Shadow Government' in the South," *The New York Times Magazine*, January 24, 1965, pp. 65–66.

[9] *The Evening Star* (Washington, D.C.), February 11, 1965, p. A–3.

[10] Franklin A. Lindsay, "Unconventional Warfare," *Foreign Affairs*, XL (January 1962), pp. 264–74.

[11] Ira Wolfert, *American Guerrilla in the Philippines* (New York: Simon and Schuster, 1945), pp. 113–14; Pomeroy, *The Forest: A Personal Record of the Huk Guerrilla Struggle in the Philippines* (New York: International Publishers, 1963), p. 163; and Molnar, *Undergrounds*, pp. 61–66.

[12] Molnar, *Undergrounds*, p. 278.

PART IV

UNDERGROUND PSYCHOLOGICAL OPERATIONS

INTRODUCTION

Few insurgencies have been won or lost by large, decisive military battles. More commonly, insurgencies are won by a combination of military and political means. Much of the political leverage involved in such settlements is derived from effective psychological operations, which have structured the environment necessary for a political solution.

One objective of psychological operations is to create social disorganization and conditions of uncertainty. The resultant unrest and confusion are used as a cover to carry out underground operations. A characteristic of this kind of social confusion is a condition of general apathy among a large segment of the populace and an unwillingness to help either side. This indifference plays into the hands of the underground: apathetic people do not cooperate by supporting government programs and they seldom volunteer the intelligence information necessary for detecting underground elements and operations. A second objective of psychological operations is the creation of doubt and suspicion of government and government officials. This focuses attention and grievances on the ineffectiveness of government. A third and crucial objective of psychological operations is to crystallize attitudes and organize dissident elements to resist government action and policies.

To insurgents, and especially Communist insurgents, influencing opinions and attitudes is not an end in itself, but only a means to enhance their organizational work among broad elements of society. The Communists state candidly that they propagandize in order to expand the mass organizations attached to their insurgent movements. Once an individual commits himself to an organization, no matter how superficial his motives or how temporary his intentions, his perceptions change and, with them, his psychological receptivity. Ultimately, it is through mass organizations that attitudes are crystallized in favor of the insurgent movement.

Psychological operations deal not only with the "objective" world surrounding a person, but with the world as seen by the individual. The "real" world or the facts are relatively unimportant in psychological operations: what people believe or can be made to believe is the important thing.

Underground propagandists and agitators identify their appeals with a society's recognized, accepted values—such as "independence" and "land for the landless." Those who accept these widely held values are led into accepting the insurgency. Insurgents also offer rewards to those who are "loyal," and threaten physical reprisals against any who oppose them. Riots and passive resistance provide strong social coercion to influence the undecided or uncommitted.

Underground psychological operations are conducted in a number and variety of forms: mass media and face-to-face persuasion; leaflets and theatrical performances; programs for local civic improvement; and threats, coercion, and terror. Although the substantive content of psychological operations during any phase is likely to be determined at the highest echelon of the organization, successful implementation depends in large part upon the ingenuity of the operators at the local level.

In attempting to influence mass action and to develop mass support, psychological operations are directed primarily to specific audiences or target groups. Occupational, religious, ethnic, and other social groups are often singled out as target groups, and tactics are tailored to be effective within a particular group. The purpose of underground propaganda may be to win support among the neutral and uncommitted; to raise morale and reinforce existing attitudes and beliefs among underground members and their supporters; to undermine confidence in the existing government; and to lower the morale of government forces and personnel.

Other psychological operations, such as terrorism, are also applied to both opponents and neutrals, to coerce them into the movement or to make them refrain from assisting the government. The underground has often killed as many neutrals to discourage collaboration with the government as they have killed members of the counterinsurgency forces. Target groups vary in reaction and must be chosen carefully. Terrorist acts by the FLN in Algeria discouraged Muslims from supporting the French, but actually increased French determination to fight. Terrorism also insures the adherence of insurgent members to discipline.

Underground movements aim different appeals at various segments of society. Groups that are reluctant to take up arms against the government can be rallied around emotional issues and directed into passive measures. Religious or pacifist groups, women and children, or old men, can be mobilized for passive resistance. In organizing demonstrations and riots, attention is given to selecting groups most likely to respond to the agitator's call to action—student groups, dissatisfied labor union members, and groups with known grievances.

CHAPTER 9

PROPAGANDA AND AGITATION

The Bolshevik Revolution added a new dimension to the concept of propaganda with the term "agitprop"—the combination of propaganda with agitation. To the Bolsheviks, "propaganda" referred to the dissemination of many ideas to a few people. Marxist-Leninist ideological propaganda was meant exclusively for the cadre, to provide them with instruction for carrying out their tasks and with inspiration to refurbish their morale; it was not intended for the masses. "Agitation," on the other hand, meant disseminating a few ideas to many people. It was believed that the masses did not understand complex issues, but could be aroused from their apathy by the constant repetition of simple, emotional issues which directly reflected their daily frustrations and needs.

In the practice of psychological operations today, however, the classic distinction between agitation and propaganda is less clear. Propaganda has come to be identified as themes and messages disseminated via mass media to a large audience while agitation has taken on the characteristic of face-to-face communications directed toward small selected audiences. These means of propaganda and agitation are interdependent and complementary. Typically, mass communications stresses the broader message of the movement and agitation assumes the task of translating this message and tying it to the grievances of specific people in specific situations.

PROPAGANDA THEMES

Propaganda is directed to the underground itself, the uncommitted, and the government. The target groups selected represent various identifiable segments of society. Persuasive themes attempt to create feelings of doubt and uncertainty about future events and to promote the feeling of crisis. Care is taken in propaganda to differentiate between the government and the people; and blame or fault is attributed to the actions of government. Among the themes are appeals to self-interest and specific needs and grievances, stressing local factors and conditions.

General propaganda themes are developed and adapted to specific purposes during a Communist dominated insurgent movement. For example, if the overall objective of Communist efforts in South Vietnam—namely, the takeover of the government by military and political means—is kept in mind, it can be noted that the themes used at the onset of the insurgency are still employed, although they have been constantly adapted to new developments.[1]

The content of persuasive messages is more surprising in what is left out than in what is included. Very little content is devoted to the ideological basis of the cause. Exaggerations tailored to strengthen emotional commitment of members are recurrent. Several devices are used to justify the movement through a "consensual validation"—creating the appearance of majority approval.

POWER AND LEGITIMACY

All propaganda, of course, stresses the legitimacy as well as the reality of insurgent power. Thus, the Viet Cong claim to be the spokesmen of the people, in contrast to the Saigon government which is characterized as "a corrupt mouthpiece of Western imperialists and neocolonialists." One of the principal thrusts of Viet Cong propaganda has been to present U.S. involvement as ruthless, unprincipled aggression. The Viet Cong capitalize upon the natural fear and ignorance of outsiders to develop plausible exaggerations about U.S. intervention. The United States is shown as obstructing the expression of the popular will—represented by the Viet Cong—and its physical presence in Vietnam proves its aggressive intent. As an aggressor, the propaganda runs, the United States will stop at nothing to achieve its objective of domination. U.S. activities are charatcerized as "atrocities," "inhuman," "brutal," and "malicious." Americans are depicted as treacherous, betraying even their staunch friends, such as Diem, whose fate awaits other "running dogs of the imperialists."

They are also described as having contempt for the Vietnamese people. To show U.S. contempt for Vietnamese religious values, the story has been widely told of U.S. planes attacking a Buddhist monastery and killing 35; and then, with fine impartiality, attacking a Catholic church and shooting 200 nuns.

Another theme of Viet Cong propaganda is that the United States is attempting to camouflage its aggressive designs and legitimize them within the world community by dragging reluctant allies and "lackeys" into the Vietnam war. Involvement of international organizations has been dismissed as a mask behind which the United States might hide its aggression.

Finally, Viet Cong propaganda points out that although the United States seems powerful, there is really no need to fear it; its forces are constantly being defeated by the Liberation Army. Constant Viet Cong victories are reported even if they must be invented, and real or purported enemy losses and desertions are emphasized to suggest an irresistible trend toward victory. New U.S. tactics will fail as certainly as past ones did. In short, say the propagandists, the U.S. efforts are the "last gasp of a dying conspiracy," so there is neither a moral nor a practical reason to support the losing and desperate Americans and their "puppets."

Nor is the Saigon regime to be taken seriously, say the Viet Cong. It is only the weak mouthpiece of the imperialistic United States. Again the Communists demonstrate their flexibility; after attacking the Diem government for years, when that government fell, the line quickly switched to point out that the "new puppets" were even more pro-American than Diem.

In sum, the Viet Cong propagandists paint lurid alternative choices for the South Vietnamese people. On the one hand is the ineffectual, U.S.-dominated Saigon government. The United States itself is weak and is suffering constant defeats. On the other hand, the Viet Cong stands for the people, for humane values, and for religion. Finally, Viet Cong victory is assured by historical necessity. What is there for the people to do? The propagandists' answer is clear: they can accelerate the process of historical determinism. Their actions, under the guidance of the Communists, can influence events and speed up the change from the present civil war and colonialism to peace and happiness. The propagandists suggest that the people can act in ways which are not only "patriotic," but also serve their self-interest. For instance, youths are urged to resist being drafted into the "losing" government army.

The Communists in South Vietnam, as elsewhere, try to increase the appearance of legitimacy by emphasizing the external support received for their cause. There is continual insistence that the insurgent movement has the support of foreign peoples even if it does not have the recognition of their governments. To present this facade, propagandists frequently misquote, cite out of context, or use isolated minority statements from non-Communist countries as expressions of mass support for the insurgents.

LOCAL APPEALS

A chief focus of underground propaganda is the local population. A study of propaganda activity in the Korean war and the Malayan and Philippine insurgencies indicates that government forces largely aim their propaganda at enemy fighters, while the insurgents seek to influence local civilians.[2]

The experiences in these three countries indicate that the government, because of the composition and approach of its propaganda effort, is often unable to maintain continuous contact with the local populace. Rarely does a government effectively influence the local population or help the people defend themselves against underground agitators.[3]

The underground, on the other hand, focus most of their propaganda effort on the local populace. The insurgents' greatest vulnerability lies in their dependence upon the cooperation of the population.[4] If the government turns the people who tacitly support the underground against the insurgents, it can decisively affect both the underground's material support and morale.

Emotional Appeals

Underground propaganda usually emphasizes emotional arguments in expressing appeals that involve local civilians, avoiding theoretical arguments or concrete statements of programs of action. The underground also restructures the situation by getting those who cannot be convinced of the rightness of the movement to see it in terms of who will win and who will lose rather than whether in the long run they are right or wrong.

Emotional appeals, outweighing the rational ones, tend to reinforce the bandwagon effect.

Threats

Threats also have been effective techniques of underground propaganda. Insurgents have been able to frighten large numbers of citizens into submission, cooperation, or at least passivity. Most often this has been achieved through recurrent threats of violence.[5]

Appeals to Self-Interest and Specific Needs

Communist propaganda often caters to the self-interest of the local population. By promising more food and better clothing and housing, they appeal to those disaffected from the government as well as to the economically deprived. When many people are unemployed, as they were in Korea, for example, the Communists find a ready audience.[6]

In Malaya and the Philippines, the insurgents were reported to have used a wider variety of topics in their propaganda than did the government forces. Rather than giving ideological reasons for the support they requested, they offered tangible advantages, such as more land, more food, and other material acquisitions to their supporters.[7]

Exploitation of Prejudices

Another example of Communist adaptation to local conditions is manipulation of long-established prejudices. In Malaya, propaganda exploited ethnic problems: the Malayans were provoked against the Tamils, an ethnic minority group, while the Chinese, who provided the chief support for the guerrillas, were depicted as pro-Malayan and benefactors of the poor.[8]

Action Propaganda

During periods of uncertainty, people tend to be suspicious of what they hear and rely on their personal experiences, believing only those things which they have seen with their own eyes. When action immediately follows promises, this is called action propaganda. There are two forms of action: one focuses on specific actions which alleviate hunger and suffering among the people and demonstrate the insurgents' ability to accomplish set goals, while the other focuses on military acts, violence, sabotage, and punishment of traitors among the local population. Both show that the insurgents are powerful in spite of being outnumbered.

In Korea, the Communist Party gave conspicuous assistance to impoverished families when no government aid was forthcoming. Cooperative food stores were started in the Philippines and in Korea; community pools of money and goods were set up in both these countries and in Malaya. Villagers in Korea frequently were abducted, exposed to propaganda, and returned the next day well-fed and unharmed.[9]

A directive from the North Korean guerrilla bureau argues that words alone cannot persuade the masses that their lot will improve under Communism; therefore, deeds, however small, are needed to convince the people of the superiority of the Communist system.[10]

The underground makes use of such mass media as leaflets, newspapers, and radio, mainly because of the rapidity with which the information can be disseminated and the size of the audience it can reach.

LEAFLETS AND NEWSPAPERS

Leaflets and newspapers are both important tools of propaganda, though they have different functions. Leaflet messages can be produced quickly to reflect sudden developments and one of the foremost tasks of an illegal party is to utilize every occasion which strikes the public imagination to interpret events to its own advantage.[11] Lenin recognized this and called for a wide distribution of leaflets. Widely scattered printed material also has the advantage that the opposing forces find it difficult to trace the source.

In underground operations, when the content of the propaganda message is geared to individual interests, the techniques of dissemination usually stress personal contact. In the Philippines, Malaya, and Korea, leaflets were passed surreptitiously from person to person by hand or by chain letter. Giving a leaflet thus implied a proof of confidence, an honor, and a privilege. It was reported that people saw, read, and remembered more of the leaflets handed to them personally than those received by mass distribution.[12]

An experiment was conducted in six Korean villages. Progovernment leaflets were dropped by air and Communist propaganda had been circulated personally. It was found that the villagers had a more accurate memory of the wording of the Communist leaflets than of government messages.[13]

Clandestine newspapers are a classical means of spreading the underground movement's message to its adherents and potential supporters. They are for many people the voice of the movement, presented in the most favorable terms. But the importance of newspapers extends beyond information and propaganda. They can become a rallying point for and even lead to the formation of an underground movement.[14] Lenin's *Iskra* is reported to have performed this function in Russia. This paper, whose distribution ranged between 4 and 15 thousand copies, held together the party which led the revolution in 1905.

Clandestine newspapers must be organized so as to provide security for the personnel involved in publication. Typically, this involves a high degree of compartmentalization and division of labor. Functions are divided and members of the staff may work separately without knowledge of other members. Editorial units draft the news and editorial articles. Couriers pick up the copy and deliver it to a composition team. Another courier delivers it to the printers. There may be a separate staff to collect funds and keep books, while a supply group obtains paper, ink, lead, etc. Thus, if one part of the publication staff is arrested, the other units are not necessarily compromised.[15]

The critical task of distribution can be carried out in various ways, depending on the local situation. Volunteers may distribute the papers individually or the regular mail may be used. If control measures are strict, various disguises can be used. For example, the Viet Cong has used the government's own materials to disguise their propaganda. In a number of villages, security forces have found Viet Cong propaganda booklets with the same covers as government booklets.[16] During World War II, the Belgian underground also used a variety of disguises for their propaganda.* One method of disguising the transmission of propaganda materials into a target country is through the diplomatic channels of embassies friendly to the underground cause. In October 1962, the Chilean Government intercepted a crate which had been shipped by the Cuban Government to its embassy in Santiago. The 1,800-pound crate was labeled "samples of Cuban products and cultural and commercial material" but actually contained subversive propaganda addressed to various Chileans who had visited Cuba within the previous few months.[19]

In the years just before the Bolshevik Revolution, the transportation of revolutionary material into Russia had been made exceedingly difficult. When the traditional ruse of the double-bottomed trunk became too risky, a method was devised whereby thin sheets of printed paper could be glued together to form boards, which were then made into cases, cartons, bookbindings, and backs of pictures. With this method, false bottoms were no longer needed because the trunks themselves could be constructed of revolutionary literature. The recipient needed only a bowl of warm water to unstick the special glue and separate the papers.[20]

Lenin claimed that one central newspaper is more efficient than a number of local ones.[21] Local news can be distributed through leaflets, but a central newspaper is a necessity. An underground press can communicate tactics and instructions in relation to specific targets and enemies. Furthermore, newspapers can serve as a training ground in which members of the organization can practice gathering and distributing information; they also learn to estimate the effects of political events on various sections of the population and to devise suitable methods to influence these events through the revolutionary party. In Lenin's words: "Arranging for and organizing the speedy and proper delivery of literature, leaflets, proclamations, etc., training a network of agents for this purpose, means performing the *greater* part of the work of preparing for future demonstrations or an uprising."[22]

Underground newspapers can help to immunize the population against official propaganda, counteract fear, defeatism, and indifference, and maintain an uncompromising hostility to the regime.[23] The printed medium gives more information and more ideas than even good oral propaganda.

---

* A small booklet calling for people to resist the Nazi occupation and giving instructions on how to slow up production might have a railroad timetable as a cover; a book of sabotage instructions might bear a pro-German-sounding title, such as *The Duty of Each in the New Order*.[17] A false edition of the German-controlled newspaper *Soir* was published by the Belgian underground. The false *Soir* reported German admissions of defeat and failures and ridiculed Hitler and the German generals.[18]

It spreads information and ideas uniformly and so is instrumental in tying its readers into a close mental community.

Undergrounds not only produce illegal newspapers, but also use existing legitimate newspapers for their own ends. In Kenya, during the Mau Mau insurrection, press criticism of certain aspects of local authority provided fertile ground for implantation of rumors about public officials. With the press background, rumors that a particular police official was "being investigated" were easily believed; accusations of police brutality or ill-treatment of prisoners were made against officials who were learning too much about or causing excessive difficulty to the Mau Mau organization.[24]

RADIO

Radio broadcasts are widely used by contemporary undergrounds and have a number of advantages over printed materials. A broadcast can simultaneously reach a large number of persons over a considerable range of territory. The same coverage by newspaper takes longer and is much more dangerous for the publishers, distributors, and recipients. A radio broadcast used in conjunction with an underground movement's agitation operations can enhance the credibility of the message and lend an appearance of strength to the organization.

An added advantage of radio is that broadcasts can be made from abroad, as in the case of Algerian FLN broadcasts over Radio Cairo and Radio Damascus. The Communist Greek underground operated Radio Station Free Greece which pretended to broadcast from within Greece, although it probably was located in Albania or Rumania.[25] Latin-American subversive groups such as the Dominican Liberation Movement, the Peruvian Anti-Imperialist Struggle Movement, and the Guatemalan Information Committee transmit hostile propaganda against their respective governments through the facilities of Radio Havana.[26]

In its early period, the Viet Cong operated a weak transmitter in the South, apparently from a junk. Subsequently, they used a much stronger transmitter for Liberation Radio to broadcast news features and commentaries in five languages: Vietnamese, English, Cambodian, French, and Chinese.[27] Radio Hanoi lent official support to the insurgents.

PROPAGANDA CONTENT

Messages transmitted by mass media during an insurgency essentially have three types of content: attention-getting, instructional, and persuasive. When people are under stress, they avidly seek information bearing upon the current crisis. As one way of capturing people's attention, underground newspapers and clandestine radio broadcasts usually contain news and information about the progress of the insurgency, what others are saying about the insurgency, events external to the conflict, and other newsworthy events. People will listen or read communications in order to get the news, but in the process they are exposed to the persuasive content of the insurgent's message.

In order to provide continuity and uniformity, selected themes are assigned and transmitted at regular intervals to all propaganda organs of the movement. Many of these instructions are transmitted undisguised through clandestine radio broadcasts. In addition, information about potential collaborators and government tactics are also contained in material distributed through mass media.[28]

The Communists have been particularly effective in the use of the "double language" routine, in which propaganda materials simultaneously give instructions to the Communist cadre. This technique had been employed in different ways. During the early phase of the Bolshevik Revolution, only indirect language was used to call for the overthrow of the government. Revolutionary propaganda resorted to "hints" which were understandable, through previous training and indoctrination, to party members or sympathizers, who in turn explained them to the people.[29]

More frequently, the Communists have used this double language propaganda technique to disguise specific instructions to the cadre. For example, one observer of the South Vietnamese insurgency has reported that it is difficult, when sorting through a pile of captured Communist documents, to say which are meant as general propaganda and which as training or instructional materials for internal use.[30]

When the Viet Cong want to pass instructions rapidly to the cadre about policy changes or a new propaganda line, they usually avoid written directives, which may take a week or more, and employ instead hidden instructions in Radio Liberation propaganda broadcasts. Since such broadcasts are overheard by the government, it is essential to make instructions appear routine and inconspicuous. Part of the cadre's training is learning to interpret camouflaged messages and to apply them to local circumstances.[31]

The persuasive messages of mass media are designed to bolster the morale of the insurgents, to undermine confidence in the government and its policies among a mass audience, and to win active supporters or at least sympathizers to the movement. The messages are phrased in highly emotional terms and may make use of distortion or complete fabrication.

Distrust of government officials is stressed and allusions to government manipulation and dishonesty are made. Other themes bring out the helplessness of individuals and the strength of the insurgent movement. A recurrent theme is that the country has an abundance of material goods for everyone but that the listeners are being excluded from their just share. Themes which stress anxiety are also included, along with warnings of disaster unless immediate action is taken.[32]

INTERNAL COMMUNICATIONS

Insurgencies in rural, agrarian, low-literacy societies face many obstacles both physical and psychological, to the free flow of communications: the absence of electricity limits the use of radios, illiteracy limits the use of newspapers, inadequate transportation routes limit travel, and so on.

144

Word-of-mouth communication has, in most of these areas, been the traditional means of contact with the outside world. Recently, forms of mass media have, to some extent, been incorporated within the traditional channels. News from the urban areas may reach a remote village through a person who has a portable radio or acquires a newspaper. This person then becomes the source of information to the entire village. Travelers—itinerant storytellers or peddlers—are another source of outside news. Institutions such as the village bazaar have taken on the function of disseminating information first planted by mass media.[33]

Word-of-mouth is an efficient means of communicating factual news of particular events, as a survey taken in 198 villages in India illustrates. Several months after the 1962 Chinese-Indian border fighting, the survey revealed that of the 83 percent of the villagers who had heard the news, 40 percent heard it from friends, shopkeepers, or similar sources. This is compared with a similar study made in the United States where 93 percent of the people surveyed had heard of particular news events within 24 hours; 88 percent of these heard the news through the mass media.[34]

The substantive and interpretive news of the journals and radio reports are usually effective only within the frame of reference of urbanized audiences. Therefore, the persons who pass news to the villages have an opportunity to add interpretation to the news they pass on by word-of-mouth. In an insurgency situation, the underground agitator seeks to take advantage of the communication process and passes on his own news with his own interpretations. His news may come from an authentic mass-media source—*Pravda*, Radio Hanoi, Radio Free Cuba, or whatever source is supporting the movement—and his interpretations are fitted to the attitudes and level of his audience and the local situation.

The agitator, to insure a maximum degree of credibility, tries to bar other external sources of information from his target group or area. This is done through the confiscation of radios, threats to rural newspaper distributors, and impairment of government access to the area. In effect, he seeks to develop a captive audience in order to facilitate his job of influencing attitudes and behavior and generating popular support for the underground movement.

AGITATION

Agitation is essential in creating mass support, for it takes more than a presentation of information to a group of people for them to accept an idea and be persuaded to it. Exposure to information does not imply absorption of it. There are psychological as well as physical barriers which inhibit the flow of information and ideas. There may be general apathy in which a large portion of the population is unfamiliar and unconcerned with particular events—the chronic "know-nothing" group. Another barrier is the phenomenon of "selective exposure": a tendency to hear only information congenial to individual tastes, biases, and existing attitudes.

There is also "selective interpretation": information understood only in terms of prior attitudes. Frequently, only differential changes in attitude appear, in which individuals who do alter their attitudes as a result of information do so only in terms of their prior attitudes.[35]

The task of the agitator lies in overcoming these barriers and putting over a message that is credible and meaningful. He must reach the indifferent; he must blend his theme with the existing attitudes of his target group; and he must make the resultant attitude one which can be converted to mass action. The agitator must also dislodge any complacency that exists among a group of people, intensify their unrest, and channel it to suit the purposes of the underground.[36]

ARMED PROPAGANDA

The concept of armed propaganda illustrates the integrated nature of military and political operations in an insurgency and stresses the relative importance of psychological and political operations. According to General Vo Nguyen Giap's view, the first of three phases of insurgent conflict is the psychological warfare phase. In the second phase, propaganda and agitation continue but armed struggle comes to the forefront. In the final phase, when victory is near, the emphasis returns to political propaganda. Before, during, and after the armed conflict, military units are charged with carrying out propaganda missions. Likewise, the political arm of the movement has armed enforcing units to support its psychological and military missions.[37]

In Vietnam, Viet Cong agitation is carried out by armed propaganda teams and cadre agitprop agents, supplemented by relatively simple village propaganda subcommittees.[38] A Viet Cong indoctrination booklet succinctly describes the nonmilitary responsibilities of its soldiers in a political struggle:

> Because of their prestige, the members of the armed forces have great propaganda potentiality. If the fighter with a rifle in his hand knows how to make propaganda, to praise the political struggle and to educate the masses about their duty of making the political attack, his influence may be very great. But if he simply calls on the population to join him in the armed struggle he will cause great damage. He must say, "Those who do political struggling are as important as we who fight with rifles. If you do not take up the political struggle we will be unable to defeat the enemy with our rifles." This will make our fellow countrymen more enthusiastic and will also help to promote the political struggle.[39]

Armed propaganda companies serving each province, although primarily military, are also responsible for psychological warfare. These units (such as those of the NFL in Vietnam in 1960) infiltrate by night those villages which by day are controlled by government troops. They may sometimes even enter during the day, assemble the villagers, make their appeals, and leave before government troops can respond. The armed propaganda units range in size from a squad to a company. Their duties include agitation, recruitment, and selective terrorism through enforcement of death sentences. They make periodic visits to villages to carry out propaganda sessions. When visiting a village, the armed unit may suggest that all villagers attend the meeting and will usually ostensibly excuse one or two reluctant persons. However, if many of the villagers fail to attend, the team may find it necessary—in the name of patriotism, of course—to enforce attendance.[40]

The possession of weapons gives such teams an air of prestige before a village audience. As a leading figure in the Philippine Huk movement stated: "The people are always impressed by the arms, not out of fear but out of a feeling of strength. We get up before the people then, backed by our arms, and give them the message of the struggle. It is never difficult after that."[41]

The tactics of armed propaganda teams vary, according to the local situation, from a simple display of armed strength to disemboweling an uncooperative village chief. In one operation in Vietnam in January 1962, 100 youth group leaders were captured by a Viet Cong armed propaganda unit. Most were released several weeks later after a period of intensive political indoctrination. Seven incorrigibles, however, were held back and probably murdered.[42]

TECHNIQUES OF AGITATION

The methods involved in armed propaganda can be illustrated by describing a typical visit of an agitprop team to a village in Vietnam.[43] Such a team includes a young man from the Youth and Student Liberation Associations; a farmer and a woman who work with the Farmer's and Women's Liberation Associations, respectively; and additional members, depending on the team's area of responsibility. Before entering the village, the team confers with an agent or sympathizer who has been in the village to investigate the people and their attitudes. If he reports that the villagers are despondent because the rice crop is poor or there are cases of cholera, this is noted as a vulnerable point.

Around dusk the team parades into the village, attracting attention by shouting and enthusiastically greeting old friends. The team suggests that everyone gather for a meeting. They are armed, so the suggestion is particularly persuasive. The team's leader tells the assembled villagers how happy the team is to see old friends, but notes that there is unhappiness in the village. To provide an informal, social aura, he may suggest that they all begin by singing old Vietnamese songs. This provides an emotional setting for the team's presentation. After a few songs the leader suggests that the people try new lyrics, dealing with the liberation of the country from imperialism. By having the audience learn and sing the new lyrics, the team conveys its message and has it repeated by the group.

The team leader then brings up the specific problems of poor crops and disease. He announces that the Americans are in the country destroying crops by chemicals—a process called defoliation. He explains that although this isn't being conducted in the immediate area, the wind carries the hateful chemicals great distances. He goes on to explain that what the villagers think is cholera actually is a rare, incurable disease resulting from American germ warfare. He concludes by asserting that if the people want better crops and better health, they must drive out the Americans.

After the leader has delivered his message, the team sings more songs and acts out a lively and humorous skit, to revitalize the audience's attention and create a stereotype of the enemy. In the skit a swaggering and boastful American makes indecent advances to an innocent young Vietnamese girl on a street in Saigon. The American is then bested by a capable, witty, and loyal Vietnamese taxicab driver.

The team, to reinforce its message, brings out a portable radio in time for the late evening news from Liberation Radio or Radio Hanoi. The broadcast tells of great victories of the National Liberation Army and the failures of American helicopters. This adds an aura of organizational strength. After the news the team gets feedback from the audience by calling for questions and comments on any subject. This will bring out specific grievances and point out potential recruits for the movement.

After the public meeting, the audience is broken up into groups—farmers, women, and teenagers. In these group discussions, the team members explain how the revolution suits their specific interests; they also determine any special grievances and make note of potential talent.

After the group sessions have broken up, the team distributes leaflets through the village, nails posters on trees, raises a National Liberation Front flag, and leaves. The next day government troops from a nearby post see the flag, come to the village, and make the people take the flag down and burn it. A few days after the team's visit, a cadre agitprop agent comes to the village and singles out those persons whom the team noted as potential recruits. He remains in the village to establish local cells or subcommittees to provide the essential follow-up and to repeat the message planted by the visiting team.

The techniques employed in such operations can be outlined:

Attention-Getting. The people's attention is aroused by loud fanfare, music, drums, displays of armed force, or any other device which creates interest and draws attention to the team's exhibition.[44]

Attention-Sustaining. People in rural areas are usually starved for entertainment and their attention can be sustained by songs, plays, motion pictures, vaudeville, magic lantern shows, etc. The agitators mix their propaganda theme with humor and entertainment by putting revolutionary lyrics to old songs and by playing skits in which the "enemy" is depicted as the villain.

Action Propaganda. Where possible, action themes rather than words or vague promises are stressed. The agitprop team makes simple promises which can be immediately implemented, so as to enhance the credibility of other propaganda themes.

Repetition. Underground "resident agents" or members of the local subcommittee set up, continue the work, and follow up the message by incorporating underground themes into everyday experiences.

Feedback. After the propaganda message has been put across, the villagers are organized into discussion groups so they can ask questions and voice complaints. The agitator thus gains insight into the specific grievances of the group and also learns the local jargon and notes activities among the villagers.

Special Interest. Agitators take advantage of social norms to add credibility to their message. They learn of special-interest grievances and adjust their theme and approach accordingly in the course of the conversation in the discussion groups.

Simplicity. The agitprop agent keeps his message simple and direct, closely related to the audience's specific interests and grievances. Later these specifics are developed into generalities.

Captive Audience. The agitator, by capitalizing on the information void in the rural areas, in effect makes the target group a captive audience.

Mass Media. Radios and newspapers are not able to carry the entire burden of persuasion or even of transmission of information in the rural areas of undeveloped nations. The agitator uses mass media mainly as a secondary, reinforcement tool.

Scapegoat. The agitator makes functional use of human tendencies to fear and reject outsiders and to transfer guilt and blame to other groups or individuals. Some distinctive physical characteristic or cultural tradition is used as a symbol at which all blame for frustration and failures is directed.[45] If the agitprop operation is successful, the target group's fears and grievances are directed toward selected outsiders such as landlords, government tax collectors, or "beak-nosed Americans."

ORGANIZATIONAL FACTORS

The Communists, in building an agitprop network, follow certain tactical principles. First, all members of the party's committee in the area of operation must be "thoroughly indoctrinated about the significance of the project before they are presented with definite plans and instructions for carrying out the program." Likewise, the local committees must be convinced of the urgency and importance of the system. The network must follow a we-do-what-we-propagate policy in order to make the message timely and stimulating.[46]

The network must seek out aggressive elements or "activists" among the target group and train them as agitators. Such people are often effective agitators because in addition to having a personal grievance they are familiar with the local people, know the attitudes, conditions, and jargon of the group, and are readily accepted.

The network is built gradually through a process of experimentation. Systems and techniques are tested in selected areas before the program is implemented generally. A propaganda plan is then drawn up at the party committee level. The plan is disseminated to local committees for further experimentation and implementation. Once the campaign is underway, its successes and failures are reviewed.

The network emphasizes one campaign at a time. Party members point out that revolution is a mass movement, that mass actions involving the population in general must concentrate on a very few definite and clearly expressed objectives. Only after one objective is completed is it replaced with another. Likewise, only after one general slogan has been learned does a second one appear.

The slogans must be combined with actual tasks in order to be meaningful and understandable to the people. In the Chinese "Resist-America Aid-Korea" campaign, agitators were supposed to make the people understand why they should hate America and what they could do individually to help oppose American "aggression." This meant such concrete actions as increased individual production or specific donations of money to buy airplanes and artillery.[47]

Communists try to transform the sentiments of the masses into an idea which appears to represent what the people want, while actually representing what the party wants. The ideas of the party must sound as though they came from the people.

The following outline is an example of a Communist propaganda plan. It was prepared for use in a small village in China:

A. *Objectives and Requirements of Propaganda*

1. There are 16 families in the section. Each of us will take care of 8 families. We both guarantee that every member of the 16 families will receive constantly the education of the Resist-America Aid-Korea Movement. At the end of half a month, we will compare notes to see which one of us is doing a better job of propaganda.

2. In addition to the fixed objectives of propaganda (the 16 families), we will talk to anyone whom we meet. The motto is not to waste one single minute or ignore one single individual. We should change the "conversation on personal affairs" into a "conversation on current affairs" and thus develop the habit of carrying on propaganda at all times and places.

3. The general task of the propaganda in the Resist-America Aid-Korea campaign should be united with the propaganda of the actual tasks carried on in the community. In the patriotic movement of increasing production, we will not only set up our own plans of production but will also mobilize all the people in the community to do the same. We will aim at mobilizing people to plant 40 acres of cotton and 60 trees and to invest in 30 shares of the local co-op; persuading 55 people to sign the Peace Appeal (Stockholm) and vote in the movement for solving the problem of Japan by a united effort (as opposed to the Peace Treaty signed with Japan by the United States and most of the other belligerents of San Francisco); organizing 30 people to participate in the demonstration parade in celebration of May 1; and directing the masses to do a good job in suppressing the counter-revolutionaries.

B. *Content of Propaganda*

1. To make everyone in the community understand that to oppose America and aid Korea is the only way to protect his home and defend the country; that the actual task of the Resist-America Aid-Korea Movement is to increase production and do a good job in one's own field; that the Chinese and Korean armies will definitely win the war, and that the American devils will eventually be defeated. Meanwhile we would point out the possible difficulties that may be encountered, explain

the experiences in China's War of Liberation, and enable the masses to understand correctly the victorious situation at present and not to be disturbed by temporary setbacks.

2. To speed up production and organize the masses of the people to participate in the movement through these principles.

3. To propagate current information on the suppression of counter-revolutionaries on the basis of the "Law on the Punishment of Counter-revolutionaries" recently made public by the government.

4. To propagate the advantages of a close relation between co-ops and the people and thus encourage people to purchase shares.

5. To propagate the meaning of signing the World Peace Appeal and voting on the Japanese question, and to explain the reasons for participating in the demonstration parade on May 1.

C. *Source of Material for Propaganda*

1. To attend the meetings for propagandists punctually, listen carefully to the lectures, and study the propaganda materials.

2. To read newspapers, propaganda handbooks, and any other material handed down from the higher Party organization and to keep constantly in touch with the secretary of the Party branch.

3. To gather reactions from the masses.

4. To maintain constant contact with the ch'u committee of the Party through letters and in person.

D. *Forms and Methods of Propaganda*

1. To organize four group discussions during this month. At least one of them should be a discussion meeting of women.

2. To conduct individual propaganda or informal conversation at least twice a day and make it a habit to do so.

3. To organize a newspaper-reading group, and read the *Ta Chung Pao* (*The Daily of the Masses*, published in North Kiangsu) every three days. We will take turns in reading newspapers.

4. To put out a "propaganda bulletin board" on current affairs and local news. The board is to be supplied with new material every three days.

5. To grasp every opportunity for propaganda such as working, walking, etc.

6. To make use of the aggressive activists in the masses. It is our plan to make use of Tsu Chang-yu [name of a child] to carry on propaganda among the 18 children in the community. We are planning to educate and use Siao Chi-yuan [name of a woman] to carry on propaganda among the 12 women in the section.[48]

AGITATION OPERATIONS

Province-level agitation units conduct workshops on general themes to be disseminated through the lower echelon districts and cells. The cadres then conduct surveys in their precincts and districts and adapt the themes to the local conditions of their areas.[49] In late 1964, there were estimated to be 4,000 Viet Cong trained agitprop cadres working in the rural areas of South Vietnam.[50]

The goal of agitprop operations is to transform grievances into power, to develop mass support which can be mobilized into mass action. The four phases in these operations are investigation and infiltration, involvement, crystallization, and commitment.

The agitprop operation begins with an investigation phase, designed to familiarize the cadre with actual conditions among both enemies and allies. Mao Tse-tung's formula is "investigate first, propagandize next." Mao emphasizes this by stating:

> Without investigation no comrade has the right to speak. . . . Take our propaganda work for instance. If we do not understand the real propaganda situation of the enemy, allies, and ourselves, we cannot correctly determine our propaganda plan. In all tasks of every department we should first have an understanding of the existing conditions so that we can do our work well.[51]

In Vietnam the normal pattern is for an agitprop team (5 to 10 men) to penetrate an area or village to investigate the local psychological and administrative strengths and vulnerabilities. The underground agitators live among the people, propagandize the insurgent cause, and organize local committees. Selective terrorism is carried out against those who fail to respond to the team's overtures.[52]

A captured Communist directive to North Korean agitators operating in the U.N. controlled city of Seoul outlined their duties, which included organizing and reviving cells within the city, keeping close check on the attitudes of the population within each precinct, and recording all incidents of South Korean civilian cooperation with the U.S. forces.[53] The purpose of this was to identify all who were friendly or unfriendly to the government.

The investigation stage, then, determines what the local tasks are, who the friendly activists are, and what the attitudes and sentiments of the people are. According to a captured Viet Cong province propaganda chief: "You must be patient in listening to people's problems. You must know what they are talking about and, most of all, you must leave the impression that their specific grievance is your main concern. With peasants you discuss land reform and maybe you promise education to youths. You don't go far on generalities."[54] Armed with inside information on the particular grievances of a village and the individuals in it, an agitprop team can prepare a program.[55]

In areas where there is little or no government control, agitprop teams can move relatively freely. Even where there is substantial government control the teams can often gain access to villages. In Malaya, for example, the Communists operated in bands of 100 or so and could overwhelm the small police force (usually a sergeant and 10 men) in Malayan villages. The Communists usually murdered all government officials and any known supporters in the village. With this background, the Communists established secret agreements with many of the rural police posts. If the police would permit the Communists to have the village at night, they could continue to run the village during the day. Mock raids were staged so that the police could drive the Communists off; thus the operations were covered up and the police even won awards. When the people of the village became aware of the deal, which they soon did, they too cooperated with the Communists.[56]

Involvement

Once an area has been infiltrated, efforts are focused on getting individuals to involve themselves with the underground movement through some overt act. At first, all that is needed is any symbolic act, such as flying a flag, cutting a single strand of barbed wire in a strategic hamlet fence, or constructing a naily board to be used against government soldiers. Once a villager is involved even superficially, it is comparatively easy to expand his activities into full-scale active support.

After he is established in an area, an agitator turns his attention to public health, sanitation, and education.[57] Humanitarian endeavors are undertaken to win sympathy and create social indebtedness. The agitators look for families who have lost everything, those who have not received needed assistance from the government, the unemployed, and the sick. Special attention is paid to the families of government soldiers and officials. A favorable word from a mother, father, or a close friend can have a greater impact than any slogan.[58]

"Action propaganda" is used to create popular support. The "actions" included helping peasants in the fields, distributing captured food to the population, repairing churches and shrines, and reopening schools.[59]

In the involvement phase the agitator also aims to confuse the opinions and emotions of his audience, so that even if they do not support the underground cause, they at least are uncertain about supporting the government.[60]

Crystallization

Traditional concepts and customary ways of thinking must be dislodged in order to create a "mobilizable" mass base for the underground movement. A successful agitprop operation must not only gain people's attention, it must provide a channel for their tensions.[61] Crystallization is the phase of ideological preparation in which the goal of the agitator is to transform grievances into hatred of the enemy, whoever he may be; to convince the people that the movement serves them and that their personal interest demands their support of the cause; to teach the people the why and how of the political struggle; and to develop faith and enthusiasm for the movement.[62]

One technique used by Communists in developing crystallization is "thought-revealing and grievance-telling" sessions. In these sessions agitators draw out and expose the mental processes of their audience. A basic rule is to have everyone talk, for talking helps a person convince himself. Since everyone has some grievance, the agitator encourages them to express their grievances to the group. Then he supplies the "correct" interpretation of the expressed grievances, coupling them with the goals of the movement.

Two basic methods are used in grievance-telling sessions. The first method is the "from far to near" approach in which the meetings deal with things in the past and then focus on present happenings. For example, the agitator may start with some esteemed national leader in history who symbolizes certain societal ideals. Slowly he ties the past symbol into the current movement—wrapping his message in the cloak of patriotism.

A second method is the "from few to many" approach: the agitator encourages a few activists in the group to initiate the grievance process and then persists until everyone rises to pour out his grievances. Each time a person offers a grievance, the agitator suggests that others have similar grievances; this involves them in the discussion and relates their grievances to current underground themes.[63]

Commitment

After the group has been investigated, infiltrated, involved, and rallied around a common motive for fighting, it must be impelled to take the desired mass action. This action may be in the form of protest demonstrations and parades or the election of a new slate of local government officials. Village youths can be recruited into military units while other villagers provide logistic support, food, intelligence, and weapons. At this "action-taking" phase the agitator truly has a captive audience.

The agitator communicates ideas and information to the people through organizations. These organizations can be old-established institutions (labor unions, rural cooperatives, etc.) which the agitators now control, or they can be newly established ones. The Viet Cong, for instance, generally prefer to create new organizations for specific motivational and control purposes.[64]

If an underground is successful in gaining control of the government, agitation operations can then develop popular loyalty and support for the new regime. People usually lose interest in a movement when their immediate grievances, such as the demand for land, have been satisfied. The agitator serves as a liaison between the controlling element and the public by passing on information and directives through his regular techniques. His new themes emphasize that apathy and opposition to the movement are wicked and hopeless.[65]

FOOTNOTES

[1] The source used for the following description of propaganda themes was the periodical published by HQ, U.S. Army, Headquarters of Broadcasting and Visual Activity, PAC, APO 331, "Communist Propaganda Trends." The periodical gives weekly and monthly summaries of Communist propaganda and shows its direction and trends.

[2] Fred H. Barton, *Salient Operation Aspects of Paramilitary Warfare in Three Asian Areas*, ORO–T–228 (Chevy Chase, Md.: Operations Research Office, 1953).

[3] *Ibid.*, pp. 29, 88.

[4] *Ibid.*, p. 3.

[5] *Ibid.*, p. 90.

[6] *Ibid.*, p. 29.

[7] *Ibid.*, p. 90.

[8] *Ibid.*, p. 43.

[9] *Ibid.*, p. 86.

[10] *Ibid.*, p. 25.

[11] Otto Bauer, *Die Illegale Partei* (Paris: Editions La Lutte Socialiste, 1939), p. 131.

[12] Barton, *Paramilitary Warfare*, p. 84.

[13] *Ibid.*, p. 27.

[14] Bauer, *Illegale Partei*, p. 128.

[15] George K. Tanham, "The Belgian Underground Movement 1940–1944" (unpublished Ph.D. dissertation, Stanford University, 1951), pp. 221–26.

[16] Seth S. King, *The New York Times* (March 17, 1965).

[17] Tanham, "Belgian Underground," p. 202.

[18] *Ibid.*, pp. 238–40.

[19] Edwin Martin, "Communist Subversion in the Western Hemisphere," *Department of State Bulletin*, March 11, 1963, p. 350.

[20] Michael Furtell, *Northern Underground* (London: Faber and Faber, 1963), pp. 39–40.

[21] V. I. Lenin, "What Is To Be Done?," *Collected Works*, Vol. 5 (Moscow: Foreign Languages Publishing House, 1961), pp. 346–529.

[22] V. I. Lenin, "Letter to a Comrade on Our Organizational Tasks," *Collected Works*, Vol. 6 (Moscow: Foreign Languages Publishing House, 1961), p. 240.

[23] Bauer, *Illegale Partei*, p. 132.

[24] Frank Kitson, *Gangs and Counter-Gangs* (London: Barrie and Rockliff, 1960), p. 46.

[25] Andrew R. Molnar, *et al.*, *Undergrounds in Insurgent, Revolutionary and Resistance Warfare* (Washington, D.C.: Special Operations Research Office, 1963), p. 307.

[26] Martin, "Communist Subversion," p. 350.

[27] Douglas Pike, "The Communication Process of the Communist Apparatus in South Vietnam," (unpublished manuscript, Center for International Studies, Massachusetts Institute of Technology, circa 1964; and unpublished Special Operations Research Office working paper), pp. 30–32.

[28] *Ibid.*, pp. 2–5.

[29] The Communist use of double language is discussed more fully in Louis F. Budenz, *The Techniques of Communism* (Chicago: Henry Regnery Co., 1954), especially pp. 41–44.

[30] Pike, "Communication Process of the Communist Apparatus," pp. 2–42.

[31] *Ibid.*

[32] Leo Lowenthal and Norbert Guterman, "Portrait of the American Agitator," *Public Opinion and Propaganda*, eds. Daniel Katz, *et al.* (New York: The Dryden Press, 1954), pp. 470–77.

[33] See Wilbur Schramm, *et al.*, *Mass Media and National Development* (Stanford, Calif.: Stanford University Press, 1964), pp. 72–87, 104; and also Max F. Millikan and Donald L. M. Blackmer (eds.), *The Emerging Nations* (Boston: Little, Brown and Co., 1961).

[34] Schramm, *Mass Media*, pp. 87–88.

[35] Herbert H. Hyman and Paul B. Sheatsley, "Some Reasons Why Information Campaigns Fail," *Public Opinion and Propaganda*, eds. Daniel Katz, *et al.* (New York: The Dryden Press, 1954), pp. 522–31.

[36] Herbert Blumer, "Collective Behavior," *Principles of Sociology*, ed. Alfred McClung Lee (New York: Barnes and Noble, 1951), pp. 202–205.

[37] General Vo Nguyen Giap, *People's War, People's Army* (Washington, D.C.: Government Printing Office, 1962), see especially pp. 76–81.

[38] Pike, "Communication Process of the Communist Apparatus," pp. 19–20.

[39] *Ibid.*, p. 39.

[40] *Ibid.*, pp. 22 and 38.

[41] William J. Pomeroy, *The Forest: A Personal Record of the Huk Guerrilla Struggle in the Philippines* (New York: International Publishers, 1963), p. 43.

[42] Bernard B. Fall, *The Two Vietnams: A Political and Military Analysis* (New York: Praeger, 1963), p. 137.

[43] Pike, "Communication Process of the Communist Apparatus," pp. 22–26.

[44] Frederick T. C. Yu, *The Strategy and Tactics of Chinese Communist Propaganda, As of 1952*, HRRI, Maxwell Air Force Base, Alabama (Lackland Air Force Base, Texas: Air Force Personnel and Training Research Center, June 1955), p. 17.

[45] Floyd L. Ruch, *Psychology and Life* (Chicago: Scott, Foresman and Co., 1958), p. 420.

[46] Yu, *Chinese Communist Propaganda*, pp. xii, 13–14; and Frederick T. C. Yu, *The Propaganda Machine in Communist China—With Special Reference to Ideology, Policy, and Regulations, As of 1952*, HRRI, Maxwell Air Force Base, Alabama (Lackland Air Force Base, Texas: Air Force Personnel and Training Research Center, 1955), pp. 29–30.

[47] *Ibid.*

[48] Yu, *Propaganda Machine*, pp. 25–26.

[49] Luis Taruc, *Born of the People* (New York: International Publishers, 1953), p. 132.

[50] Pike, "Communication Process of the Communist Apparatus," p. 22.

[51] Yu, *Chinese Communist Propaganda*, p. 14; Mao Tse-tung, "The Reform of Learning," Cheng Feng Wen Hsien (Documents of the Party's Ideological Remoulding Movement).

[52] Bernard B. Fall, "South Viet-Nam, 1956 to November 1963," (unpublished working paper, Special Operations Research Office, Project NUMISMATICS).

[53] Fred H. Barton, *North Korean Propaganda to South Koreans (Civilian and Military)*, Technical Memorandum ORO–T–10 (EUSAK) (Chevy Chase, Md.: Operations Research Office, 1951), p. 109.

[54] Stanley Karnow, "Captured Red Gives Clues to Viet-Nam War," *The Washington Post*, May 19, 1964, p. A–16; see also Taruc, *Born of the People*, pp. 37–38.

[55] Pike, "Communication Process of the Communist Apparatus," p. 23.

[56] Richard L. Clutterbuck, Speech at Special Operations Research Office on December 18, 1963; see also Richard L. Clutterbuck, "Communist Defeat in Malaya: A Case Study," *Military Review*, September 1963, pp. 63–78.

[57] Denis Warner, *The Last Confucian* (New York: Macmillan, 1963), p. 124.

[58] Molnar, *Underground*, pp. 59–60; Barton, *North Korean Propaganda*, p. 110.

[59] Alexander Dallin, Ralph Mavrogordate and Wilhelm Mill, "Partisan Psychological Warfare and Popular Attitudes," *Soviet Partisans in World War II*, ed. John A. Armstrong (Madison, Wis.: University of Wisconsin Press, 1964), p. 255.

[60] Pike, "Communication Process of the Communist Apparatus," p. 21.

[61] See Blumer, "Collective Behavior," pp. 204–205; and Yu, *Chinese Communist Propaganda*, pp. 15–16.

[62] Pike, "Communication Process of the Communist Apparatus," p. 25.

[63] Yu, *Chinese Communist Propaganda*, pp. 18–19.

[64] Pike, "Communication Process of the Communist Apparatus," p. 27; and United States Information Service, *Inventory of Communist Documents* (Saigon: U.S. Information Service, December 1, 1963).

[65] See Warner, *Last Confucian*, p. 127.

CHAPTER 10

PASSIVE RESISTANCE

Nonviolent passive resistance has long been a method and corollary of human conflict, playing a role in many underground and revolutionary activities. Although both the role and, particularly, the process of passive resistance have sometimes been obscured by philosophic and religious considerations, its tactics remain much the same, whether practiced by a Gandhi from moral and religious convictions or by a pragmatic underground as an expedient. Typically, the underground organizes and directs passive resistance techniques, and persuades the ordinary citizen to carry them out.

OBJECTIVES

Passive resistance implies a large, unarmed group whose activities capitalize upon social norms, customs, and taboos in order to provoke action by security forces that will serve to alienate large segments of public opinion from the government or its agents. If the government does not respond to the passive resisters' actions, the resisters will immobilize the processes of public order and safety and seriously challenge the writ of government.

Passive resistance rests on the basic thesis that governments and social organizations, even when they possess instruments of physical force, depend upon the voluntary assistance and cooperation of great numbers of individuals.[1] One method, therefore, of opposing an established power structure is to persuade many persons to refuse to cooperate with it.

The principal tactic used to induce noncooperation—tacit withdrawal of the populace support of the government—is frequently described as persuasion through suffering. One of the persistent myths of passive resistance is that persuasion through suffering aims only to persuade the opponent and his supporting populace by forcing him to experience a guilty change of heart and a sense of remorse: "The sight of suffering on the part of multitudes of people will melt the heart of the aggressor and induce him to desist from his course of violence."[2]

This conception of the role of suffering makes the fundamental error of presuming that only two actors are involved in the process of passive resistance: the suffering resister and the opponent. Actually, passive resistance operates within a framework involving three actors: the suffering passive resister, the opponent (the government and security forces), and the larger "audience" (the population). Every conflict situation is dramatically affected by the extent to which the audience becomes involved. One political scientist has called this phenomenon "the contagiousness of conflict."[3] A great change inevitably occurs in the nature of the conflict when the audience is included as a third actor. The original participants are apt to lose control or, at the least, the outcome is greatly influenced. It is most

157

important in politics to determine the manner and extent to which the scope of conflict influences the outcome, and how to manipulate it. Passive resistance techniques, particularly the function of suffering, provide one insight into the manipulation of "the contagiousness of conflict." More than anything else, the objective of passive resistance is to create situations that will involve public opinion and direct it against the established power structure. When this happens, the position of the passive resisters is legitimized.

When the passive resister suffers at the hand of the government, it demonstrates his integrity, commitment, and courage, while showing the injustice, cruelty, and tyranny of the government. The essential function of suffering is comparable to the interaction that takes place between a martyr and a crowd. The passive resister's token to power in the face of the security force is his capacity to suffer in the eyes of the onlooker. The courage and dedication of an unresisting martyr can have tremendous impact on the imagination of a crowd.[4]

If the passive resister provokes a response from the security forces or government which can be made to seem unjust or unfair, his charges of tyranny and persecution are confirmed. Should the government fail to act, it abdicates its control over the population, over the enforcement of law, and over the maintenance of order. Passive resistance techniques suddenly thrust upon a government the initiative, and also the responsibility, for uninvited conflict with unarmed citizens.

The primary function of passive resistance suffering is to redraw the political battlelines in favor of the resister. There are, of course, a number of variables in the effectiveness of suffering. One is the attitude and orientation of the opponent; success seems dependent on whether the opponent really cares how a population views him, whether or not he is attempting to win favor. Also, the coercive effect depends on whether or not the opponents are the passive resister's own countrymen; if they are, common identity and nationalism tend to induce empathy. In some societies passive suffering is viewed with contempt and is seen as masochism or "an exploitation of the rulers' good-natured reluctance to allow unnecessary suffering, denying attributes of personal courage or virtue to the sufferer."[5]

In addition to alienating public opinion from a government, underground-sponsored passive resistance has two other equally important objectives. The first is to lower the morale of government officials and security forces. This goal is most relevant to occupying forces. A second objective is to tie down security forces. By organizing and encouraging a citizenry to use techniques of passive resistance, the underground can successfully divert security forces from other tasks.[6]

TECHNIQUES

The arsenal of the passive resister contains a number of weapons of nonviolence. One reason they may be effective is that the government forces may not know how to cope with nonviolence. Police and soldiers are trained to fight force with force, but are usually "neither trained nor psychologically prepared to fight passive resistance." [7] Actions of passive resistance may range from small isolated challenges to specific laws, to complete disregard of governmental authority, but the techniques of passive resistance can be classified into three general types: attention-getting devices, noncooperation, and civil disobedience.

ATTENTION-GETTING DEVICES

Passive resistance in the early stages usually takes the form of actions calculated to gain attention, provide propaganda for the cause, or be a nuisance to government forces. Attention-getting devices include demonstrations, mass meetings, picketing, and the creation of symbols. Demonstrations and picketing help advertise the resistance campaign and educate the larger public to the issues at stake. Such activities provide propaganda and agitation for both internal and external consumption.

An example of this was the 1963 Buddhist protest in South Vietnam. The self-immolation of a Buddhist monk was strategically timed to insure that newsmen and photographers—particularly U.S. newsman—would be present to record the event. The leader of the Buddhists, Thich Tri Quang, wanted publicity in the U.S. press and took pains to make U.S. reporters welcome. [8]

The creation of symbols is also a common passive resistance device. Besides gaining attention, the Buddhist monks who immolated themselves—particularly the first, whose heart was preserved and displayed in Saigon's Xa Loi Pagoda—also became symbols for the resistance campaign. An example of another kind of symbol is seen in the Danish resistance movement against the Nazis. King Christian became a symbol embodying the spirit of the passive resistance struggle. The King's traditional morning ride through Copenhagen on his statuesque horse, unaccompanied by police or aides-de-camp, even months after the Nazi occupation, gained national attention. As the Danish poet, Kaj Munk, wrote then: "It does us good, as if it says to us, Denmark is still in the saddle." [9] The King also kept his royal standard flying both day and night over his palace, indicating he was always ready—either to negotiate with the Germans or lead his people. The tales of his veiled rudeness to German officials set the tone for all Denmark. A German general led a sizable force in an inexplicable attack on the weakly guarded palace, killing several Danish soldiers who put up courageous resistance. The King's sarcastic greeting to him—"Good morning, my brave general"—was told and retold throughout the nation. [10]

Jokes are often used to provoke an enemy and demonstrate contempt. This technique was used by most passive resistance movements in Nazi-occupied Europe. Sometimes it took the form of shouting anti-Nazi jokes in a cinema hall showing German films, or of little jokes made in public about Nazi repression policies, like the Danish streetcar conductor calling out, "All saboteurs change here." [11]

Ostracism campaigns, accusations, whispering campaigns, and refusal to speak or be friendly are also frequent techniques. In the anti-Nazi resistance, these occasionally developed spontaneously; later, they were often organized by the undergrounds. In Denmark these techniques were labeled *Den kolde Skulder* (the cold shoulder) policy, and many people wore buttons initialed DKS or SDU (*Smid dem ud,* or throw them out). Open contempt was displayed: "If a German military band gave a concert in a public place, they did so without a single listener. If Germans entered a cafe, at a given signal all Danes then rose and left." [12]

In Belgium similar activities were organized. One illustration of the type of witticism which helped the Belgian morale and enraged the Germans centered on the proposed German invasion of England: "An attractive housewife entered a store just before a German officer, who of course told the storekeeper to help the lady first. The lady declined and stated she did not wish to delay the officer who was probably in a hurry to catch his ship for England." [13] Also, anti-German inscriptions began to appear on the sides of buildings, on sidewalks, and on streets. In fact, one ingenious Belgian reportedly cut letters spelling "Down with the *Boche"* in the rims of his automobile tires, then filled them with paint, so that the slogan was painted continuously down the middle of the street. [14]

Such programs served as a two-edged sword: they lowered the morale of the Germans while at the same time raising the morale of the populace, creating a feeling of defiance and unity which could be channeled later into more significant resistance activity.

Nuisance activities vary greatly. They may be offensive personal acts against the opponent, such as the Algerian children publicly spitting on French soldiers. If a soldier struck a child, public opinion against the French would solidify all the more. The soldier felt humiliated and was clearly shown how the populace felt about his presence.

Another nuisance device is to overload the government security system with reports of suspicious incidents and persons. By following government instructions, large numbers of people can turn in false alarms or make unfounded denunciations of people who are suspected of aiding the enemy and in this way so overload governmental authority that valid reports cannot be handled. This technique has frequently been used against the block-warden surveillance system of countering underground activities. [15]

NONCOOPERATION

Techniques of noncooperation call for a passive resister to perform normal activities in a slightly contrived way, but not so that police or government can accuse him of breaking ordinary laws. Such activities as "slow-downs," boycotts of all kinds, and various forms of disassociation from government are all examples of noncooperation.

There are numerous examples of noncooperation in the anti-Nazi resistance movements, including falsification of blueprints and deliberate errors in adjustment of machine-tools and precision instruments. [16] Workers in shipping departments of Nazi factories addressed shipments to the wrong address or conveniently forgot to include items in the shipment. Feigned sickness was widespread. [17] These acts of noncooperation impeded the war

160

effort while appearing simply to be honest mistakes. In Yugoslavia railroad workers used a particularly effective noncooperation technique: during an Allied air raid they deserted their jobs and, after the raid, they stayed away for 24 hours or more because of "feigned fear." This seriously delayed railway traffic.[18]

Noncooperation is a principal tool of passive resistance and has been shown to be most effective in disrupting the normal processes of society and severely hampering and challenging the writ of a government—all in a way that is difficult for the government and its security forces to challenge. Many individuals altering their normal behavior only slightly can add up to a society behaving most abnormally.

CIVIL DISOBEDIENCE

Mass participation in deliberately unlawful acts—generally misdemeanors—constitutes civil disobedience. This is perhaps the most extreme weapon of passive resistance; the boundary between misdemeanors and serious crimes can be considered the dividing line betwen nonviolent and violent resistance.

Forms of civil disobedience include the breaking of specific laws, such as tax laws (nonpayment of taxes), traffic laws (disrupting traffic), and laws prohibiting meetings, publications, free speech, and so on. Civil disobedience can also take the form of certain kinds of strikes and walkouts, resignations en masse, and minor destruction of public or private property.

In Palestine, after the Haganah raided the British and hid in a nearby village, passive resistance by the Jewish population was effective in preventing their capture. When the police began a search, people vigorously refused them entrance to their homes, stopping only short of using arms; often hand-to-hand fighting with bricks and stones broke out, and first-aid stations were set up to treat the injured. At the first sign of a British cordon, a gong or siren would sound, at which signal villagers from nearby settlements would rush into the area, flooding it with "outsiders" and effectively preventing British recognition of which "outsiders" had taken refuge in the village following the raid and which had come simply to create confusion.[19]

Civil disobedience is a powerful technique, but to be effective it must be exercised by large numbers. There is a calculated risk involved: the breach of law automatically justifies and involves punishment by the government and security forces. However, the more massive the scale on which civil disobedience is organized, the less profitable it is for the government to carry out sanctions. For example, during a Huk-led strike in the Philippines, as police were attempting to arrest the leaders, Luis Taruc used the tactic of demanding and forcing the government to arrest everyone participating in the strike. "We must crowd the prison with our numbers," he said. "If there is no room for us [in the police vans], we will walk to jail." It quickly became unfeasible for the security forces to use the threat of jail.[20]

During the Indian independence movement, Gandhi effectively used the same tactic. He led so many millions in the breach of law that it proved impractical, if not impossible, for the British to jail all offenders. As British officials saw, such widespread disrespect for a law makes its enforcement ridiculous and counterproductive.[21] Yet, if a government cannot enforce its writ, it must abdicate authority. As the jails became impossibly full, Gandhi's position for pressing his demands on a government searching for ways to pacify the population was increasingly enhanced.[22]

Organizers of passive resistance are selective about the laws that are to be broken. The laws should be related in some manner to the issues being protested or the demands being made. Examples are Gandhi's selection of the salt tax in India, which was considered a hardship tax on the peasants and representative of unjust British rule; the Negro sit-ins in the United States, which are directly related to discrimination in public places; and the Norwegian teachers' strike against the Nazi puppet-government's demands that teachers join a Nazi association and that Nazi Socialism be taught in the schools. If the laws that are broken have little or nothing to do with the issues involved, it is difficult to persuade a citizenry to risk government sanctions.

In summary, the underlying consideration in most passive resistance techniques is whether they serve to legitimize the position of the passive resister while alienating or challenging the government.

ORGANIZATION

Obviously, the success of passive resistance rests largely on its ability to secure widespread compliance within the society. A government cannot be robbed of the popular support upon which it depends if only a few individuals act. A boycott, for example, requires participation by great numbers.

Organization is of critical importance to passive resistance. Although a few individuals can launch a passive resistance movement, in order to succeed they must be joined by thousands whose participation is strategically channeled. How is widespread social compliance secured? What forces and factors induce people to practice passive resistance?

NORMATIVE FACTORS

One method by which leaders of passive resistance movements secure widespread compliance is by clothing their movement and techniques in the beliefs, values, and norms of society—those things people accept without question.

For example, the earliest stages of the Norwegian resistance against the Nazi occupation and Quisling puppet government were led by the clergy. From pulpit and parsonage, the religious leadership of Norway coalesced public opinion against the Nazis by invoking the voice of the church. "When the Nazis established a new ecclesiastical leadership, the bulk of the old established Church ignored the orders of the new hierarchy. Through nonviolent action it preserved its integrity," simply refusing to cooperate in religious affairs with the Nazi occupation.[23] Because the institutions of religion were held in high esteem, and because the clergy appealed within the framework of religious values, the Nazis never were able, even in the later stages of the occupation, to break the church's resistance.

In India, women were used on the frontlines of demonstrations, making it awkward for British forces to break up the crowd without inflicting injuries on the women and further stirring up public opinion.[24]

CONSENSUAL VALIDATION

The technique of "consensual validation"—in which the simultaneous occurrence of events creates a sense of their validity—is often used to coalesce public opinion. For example, if demonstrations take place at the same time in diverse parts of a country, the cause which they uphold appears to be valid simply because a variety of persons are involved. A minority group can organize a multitude of front organizations, so that seemingly widely separated and diverse organizations simultaneously advocate the same themes and give the impression that a large body of opinion is represented. Passive resistance organizers effectively use the psychology of "consensual validation" to rally public opinion.

MYSTICAL FACTORS

Rare or extraordinary factors such as charisma play an important part in mobilizing public opinion in a passive resistance movement. Gandhi's leadership of India's independence struggle verged on the mystical. Thousands of villagers from rural India, who perhaps could not be touched nor aroused by any modern means of communication or organized population pressure, were stimulated into action by Gandhi's fasts and his religious mystique. Hundreds of thousands of peasants gathered to meet Gandhi although they often did not understand his language and could barely see him. It is difficult to estimate the role of Gandhi's mystique in coalescing public opinion against the British, but it is clear that resistance to colonial rule had never appeared on such a large scale before Gandhi.

PRESSURE FOR CONFORMITY

The same techniques used by passive resisters against the government can be used to insure widespread social compliance within the resistance movement. Ostracism is frequently used to apply pressure on individuals not participating in the passive resistance campaign. Instances of organized ostracism of collaborators can be found in all the underground resistance movements in Nazi-occupied countries during World War II. In Denmark, the underground published blacklists that were "feared by all those who acted in the interest of the enemy." [25] Informal, everyday pressures of conformity also help secure widespread compliance. The fact that passive resistance generally appears during times of crisis or of popular unrest indicates that at such times there is often a greater sense of nationalism, of a particular "we" arraigned against "they." There are then strong pressure demanding conformity.

COMMUNICATION AND PROPAGANDA

As noted earlier, the first phase of passive resistance is characterized by a period of attention-getting propaganda activities: parades, demonstrations, posters, newspapers, and other forms of communication, either clandestine or open. Once the resistance movement is launched, there must be continuing means of "spreading the word." No movement can operate without some form of communication between the leaders and the led. The underground press in Denmark played an important role; the first illegal newssheet appeared on the very first day of the German occupation.[26] Within a year, nearly 300 illegal newspapers were being published once or twice a month. They had an estimated circulation of 70,000, and each copy was read and passed on by large numbers of people. Ironically, "as the German domination became sharper, so did the free press become more and more powerful." [27]

Similar examples can be found in Norway, Poland, France, and the Netherlands during the Nazi resistance. Besides providing "objective" information on the course of the war, using Allied news-broadcast reports, the clandestine papers instructed the populace in passive resistance techniques and procedures.[28]

Clandestine methods of communication resembling those used by espionage organizations were also developed. One of the earliest and most successful acts of resistance in Norway was the dramatic resignation in 1941–42 of 90 percent of Norway's teachers, as a response to Nazi pressure to teach National Socialism and join a Nazi teachers' association. The manner in which this dramatic rebuke to the Nazi regime was carried off, and the means used to communicate the plan to all of Norway's teachers, are related in the following personal account:

> A friend telephoned me one afternoon and asked me to meet him at the railway station. There he gave me a small box of matches.
>
> He told me we teachers were to follow the lead of those [teachers] who had met secretly in Oslo, and that all the possible consequences had been discussed.
>
> The box of matches contained a statement [which all teachers were to make simultaneously to the Nazis]. My job was to circulate it secretly among the teachers in my district. That was all I knew. I didn't know who the "leaders" were who met in Oslo.[29]

Such impersonal communication was obeyed, even though the leaders were never known, because each communicator was trusted and people were assured others were going to take similar action.

TRAINING

Once organizational steps are taken to secure widespread social compliance, an effort must be made to instruct and train passive resisters. The idea is to erect a mental barbed-wire fence between resisters and authority. This instruction often takes the form of codes of "do's and don'ts." Many undergrounds have found that it is easier to tell people what not to do than what to do.[30]

Training is particularly critical when positive, not just negative, actions are desired. Noncooperation and civil disobedience are positive acts that necessarily involve training, organization, and solidarity on the part of the resisters, whether they operate in the open or clandestinely.

Gandhi placed great emphasis on nonviolent training, not only because he looked upon nonviolence as a moral creed, but because he understood that it was essential for effective passive resistance. He required his followers to swear to an oath and he developed a code for volunteers.[31] When individual suffering is involved, and when individuals must invite suffering through civil disobedience, considerable discipline is required. The American Negro civil rights movement followed the Gandhi example and applied it in planning the 1958 "sit-in" movement. Special schools were established to train young people to withstand physical violence and tolerate torment without responding with violence which might negate the entire stratagem. Any violent response on the part of the passive resisters might provide a justification for the use of violent procedures in quelling demonstrations.

PARALLEL GOVERNMENT STRUCTURE

One method that is frequently used to both undermine public confidence in a government and secure population compliance to passive resistance is the establishment of parallel structures of government. If a population must depend upon an underground-sponsored "government," it will be forced to comply with the underground's passive resistance program and de facto withdraw its support from the regular government.

In India, for example, Gandhi felt that the highest form of passive resistance would be the establishment of parallel institutions of government —not only because they would be a potential weapon against the British colonial government, but for their positive value in creating a sense of unity and community within the diverse Indian population. Indeed, many observers assert that the importance of the passive resistance movement in India was not so much what it did against the English as what it did for the Indians: It shaped a new Indian nationalism and provided an opportunity for Indians to repair the wounds in national self-esteem inflicted during more than 100 years of outside rule.[32]

Similarly, in Poland the underground passive resistance movement helped the Polish people maintain a sense of national identity and unity in the face of Nazi harshness. The establishment of a "secret state"—of underground courts, schools, and civil government—maintained a continuity and identification with nationalism, thus denying loyalty to the occupier.[33] In Algeria, passive resistance served to solidify the Arab community. Although unfavorable environmental, social, and political conditions had existed for generations, there was no insurgency until Arab grievances crystallized into national consensus. In Palestine, the insurgent-led passive resistance campaigns did as much to develop a strong feeling of unity and nationalism as they did to the British forces. The main effect was to lead people away from any form of support for the official government. Thus a consensual validation of the values and objectives of the insurgents was provided.

The techniques and societal values capitalized upon to undermine popular support of the government also serve the positive function of solidifying public opinion around a larger sense of community and national identification.

FOOTNOTES

[1] Mulford, Q. Sibley (ed.), *The Quiet Battle* (Garden City, N.Y.: Anchor Books, 1963), p. 9.

[2] Leo Kuper, *Passive Resistance in South Africa* (New Haven, Conn.: Yale University Press, 1957), p. 84.

[3] E. E. Schattschneider, *The Semi-Sovereign People* (New York: Holt, Rinehart and Winston, 1960), chapter I.

[4] C. M. Case, "The Social Significance of Non-violent Conduct," *The Quiet Battle*, ed. Mulford Q. Sibley (Garden City, N.Y.: Anchor Books, 1963), p. 57.

[5] Kuper, *Passive Resistance*, p. 85.

[6] T. Bor-Komorowski, *The Secret Army* (London: Victor Gollancz, 1950), p. 79.

[7] Feliks, Gross, *The Seizure of Political Power in a Century of Revolutions* (New York: Philosophical Library, 1958), p. 51.

[8] Denis Warner, "Vietnam's Militant Buddhists," *The Reporter* (December 3, 1964), p. 29.

[9] Lt. Col. Th. Thaulow, "King Christian X," *Denmark During the German Occupation*, ed. Borge Outze (Copenhagen: The Scandinavian Publishing Co., 1946), pp. 135–36.

[10] *Ibid.*, p. 136.

[11] Ronald Seth, *The Undaunted: The Story of the Resistance in Western Europe* (London: Frederick Muller, Ltd., 1956), pp. 99–100.

[12] *Ibid.*

[13] George K. Tanham, "The Belgian Underground Movement 1940–1944" (unpublished Ph.D. dissertation, Stanford University, 1951), p. 64.

[14] *Ibid.*, pp. 81–82.

[15] Bor-Komorowski, *Secret Army*, p. 79.

[16] G. de Benouville, *The Unknown Warriors* (New York: Simon and Schuster, 1949), p. 197.

[17] John G. Williams, "Underground Military Organization and Warfare" (unpublished M.A. thesis [thesis 452], Georgetown University, Washington, D.C., February 1950), pp. 89–90.

[18] David Martin, *Ally Betrayed* (New York: Prentice-Hall, 1946), pp. 177–79.

[19] R. D. Wilson, *Cordon and Search* (Aldershot, England: Gale and Polden, Ltd., 1949), p. 33.

[20] Luis Taruc, *Born of the People* (New York: International Publishers, 1953), pp. 41–42.

[21] India, *India in 1930–31, A Statement Prepared for Presentation to Parliament* (Calcutta: Government of India Central Publications Branch, 1932), p. 73.

[22] See Joan V. Bondurant, *Conquest of Violence: The Gandhian Philosophy of Conflict* (Princeton, N.J.: Princeton University Press, 1958), pp. 91ff.

[23] Sibley, *The Quiet Battle*, pp. 156–57.

[24] India, *India in 1930–31*, p. 660.

[25] Halfdan Lefevre, "The Illegal Press," *Denmark During the German Occupation*, ed. Borge Outze (Copenhagen: The Scandinavian Publishing Co., 1946), p. 64.

[26] Seth, *The Undaunted*, p. 108.

[27] Lefevre, "The Illegal Press," pp. 63–64.

[28] E. K. Bramstedt, *Dictatorship and Political Police: The Technique of Control by Fear* (New York: Oxford University Press, 1945), p. 210.

[29] Gene Sharp, "Tyranny Could Not Quell Them," in Sibley, *The Quiet Battle*, p. 171.

[30] Bramstedt, *Dictatorship and Political Police*, p. 210.

[31] Bondurant, *Conquest of Violence*, p. 39.

[32] Susanne Hoeber Rudolph, "The New Courage: An Essay on Gandhi's Psychology," *World Politics*, XVI, No. 1 (October 1963), pp. 98–117.

[33] Bor-Komorowski, *The Secret Army*, *passim;* and Jan Karski, *The Story of a Secret State* (London: Hodder and Stoughton, 1945), *passim.*

CHAPTER 11

TERRORISM

Few liberation or underground movements have abstained from terrorism. Insurgents have seldom relied solely on the attractiveness of their appeals or on the persuasiveness of their goals to secure popular support; they have generally assumed that people never entirely pursue idealistic goals or do what "logic" might tell them is most beneficial. Coercive means are therefore used to focus public attention on the goals and issues identified as important by the insurgents. Negative sanctions are employed to ensure that recalcitrant individuals comply and do "what is best for them." Essentially, terror* is used to support other insurgent techniques and operations, such as propaganda and agitation.

OBJECTIVES OF TERRORISM

The utility of terror for a subversive movement is multifarious: it disrupts government control of the population, demonstrates insurgent strength, attracts popular support, suppresses cooperation with the government by "collaborators" and "traitors," protects the security of the clandestine organization, and, finally, provokes counteraction by government forces.

DISRUPTION OF GOVERNMENT CONTROL

A common aim of terrorism is "the destruction of the existing organization of the population" [1]—that is, the disruption of government control over the citizenry. Terrorist acts are directed toward governmental officials and key supporters, making it unsafe to be a government official and, through systematic assassination, crippling the actual functioning of government.

The theory that subversive use of terror intimidates and disrupts government was popularized and applied as early as 1892 by the Russian terrorist, Stepniak, in the terrorist campaign against the Czarist regime. Stepniak understood that terrorism would not obtain the immediate, decisive, victory that a military battle would.

---

* Terror may be described as a state of mind. Its effect upon individuals cannot always be determined from an objective description of the terrorist act. That which threatens or terrorizes one individual may not affect another in the same way. Essentially, however, the process of terrorism can be viewed in the following manner: The stimulus is the threatening or terroristic act, and the response is the course of action, or inaction, pursued by the individual upon perceiving and interpreting the threat. If the perception of the threat leads to disorganized behavior such as hysteria or panic or the inability to take appropriate action, the individual is said to be in a state of terror.

Terror is not a static phenomenon: As threatening acts accumulate or escalate, the degree of terror heightens. A stimulus can be anything from an act of social sanction to threats of physical violence or actual physical attack. The corresponding interpretation of these threatening acts is a heightening state of terror. The response may vary from coerced compliance to acquiescence, from physical flight to psychological immobilization and breakdown.

But another victory is more probable, that of the weak against the strong . . . In a struggle against an invisible, impalpable, omnipresent enemy [the terrorist], the strong is vanquished . . . by the continuous extension of his own strength, which ultimately exhausts him . . . Terrorists cannot [immediately] overthrow the government; but having compelled it, for so many years, to neglect everything and do nothing but struggle with them, they will render its position untenable.[2]

The application of this theory can be seen in South Vietnam, where the Viet Cong have reportedly weakened government control in certain regions by killing and kidnapping province chiefs, police officials, village guards, and landlords. It has been estimated that in the 5-year period between 1959 and 1964, more than 6,000 minor Vietnamese officials were murdered by the Viet Cong.[3] Schoolteachers, social workers, and medical personnel have also been favorite targets. The allegiance of the people is the chief prize in an insurgency, and since schoolteachers, as one observer noted, "form young minds and educate them to love their country and its system of government, to close such schools or to cow the teachers into spreading antigovernment propaganda can be a more important victory than to defeat an army division."[4] Almost 80,000 South Vietnamese schoolchildren had been deprived of schools by 1960 because of terrorist action: 636 schools were closed, approximately 250 teachers were kidnapped, and another 30 were killed. The Viet Cong also disrupted the South Vietnamese social welfare and medical program; a highly successful malaria-eradication program, for example, was stopped in 1961 because of high casualties among its personnel caused by terrorists.[5]

To underscore the unprofitability of being, or becoming, a government servant, the Viet Cong carried out assassinations by unusual, brutal, or mysterious methods. Vietnamese village headmen suspected by the Viet Cong "of cooperating with the government or guilty of 'crimes against the people' were disemboweled and decapitated, and their families with them."[6] The attention value of such acts in the press and through word of mouth has been great and the implied threat to others obvious.

When, through the simple process of attrition, the machinery of government in one area comes to a virtual standstill, the Viet Cong reestablishes social order by setting up a "shadow government" with its own "officialdom" to collect taxes, operate schools, and implement population control measures.[7]

The proper selection of targets is important. Viet Cong terrorists were particularly directed to seek as first targets government authorities who were corrupt or unpopular; after murdering them they could then boast: "We have rid you of an oppressor." A captured copy of the Viet Cong "Military Plan of the Provincial Party Committee of Baria" specifically orders "squads in charge of villages and agrovilles to carry out assassination missions . . . and prime targets should be security forces and civil action district officials, *hooligans and thugs*."[8] By relieving the community of undesirables through selective assassination, Viet Cong terrorists seek to win popular support while crippling the operation of government.

170

However, terror can politically boomerang if the target is unwisely chosen or the assassination unwisely timed. An example of this occurred in the Philippines when a contingent of Huk insurgents ambushed and killed Aurora Quezon, wife of the Philippine president, along with her daughter and other distinguished citizens. Since she was widely known and respected by the people, Mrs. Quezon's death was a serious setback to the Communists. Reacting to the nation's feeling of condemnation, the Huk leaders declared that the terrorists acted without orders.[9]

DEMONSTRATION OF STRENGTH

In a terrorist campaign, the individual citizen lives under the continual threat of physical harm. If government police are unable to curb the terrorists' threats, the citizen tends to lose "confidence in the state whose inherent mission it is to guarantee his safety."[10] The use of terror, when effective, convinces the people of the movement's strength. Captured Viet Cong documents indicate that this is one of its primary objectives in South Vietnam. Through a demonstration of strength by effective assassination of government and village leaders, it attempts to "convince the rural population that the regime in Saigon cannot protect them."[11]

In order to publicize the movement's strength, some terrorist activities are conducted publicly. The FLN in Algeria used this tactic. Witnesses of terrorist acts were not eliminated but were spared in order to confirm the FLN success.[12] Muslims who supported the French were warned by letters bearing the FLN crest to desist from cooperating with the French; if a Muslim refused, the FLN execution order was attached to the victim's dead body. "By this method, the FLN silenced its opposition, weakened the . . . French by depriving them of the support of . . . Muslim leaders, and at the same time, assassination enhanced the prestige of the FLN . . . by affording tangible proof of the organization's effectiveness and intrepidity."[13]

However, in pursuing a similar tactic, the Communist insurgency in Malaya got itself in a difficult position. Unable to carry out and win a guerrilla war, the Communists attempted by mass terror to demonstrate their strength and neutralize the members of the population who were supporting the government. They soon received complaints from their political arm, the Min Yuen, that indiscriminate terror was alienating the voluntary support upon which their long-range success depended.[14]

It has been said that "terror is a psychological lever of unbelievable power—before the bodies of those whose throats have been cut and the grimacing faces of the mutilated, all capacity for resistance lapses . . . "[15] However, there is evidence to suggest that without positive inducements or popular support it may also backfire. For example, in Algeria terrorism in the name of nationalism largely won support from or cowed the Muslim population, but it had an opposite effect on the French community. Although thoroughly unnerved by the FLN terrorist offensive of 1956, they were never on the point of surrendering. In itself the dramatic nature of the terrorist challenge ensures a dramatic response to the call for counteraction. The effect of FLN terrorism on French policy was to strengthen the resolve to stamp out the rebellion, indeed to make it politically impossible to follow any other course." [16]

PUNISHMENT AND RETALIATION

The threat of punishment to collaborators and informers within the general populace is a common feature of insurgent terrorism. In fact, many insurgencies against occupying powers have taken a higher toll of indigenous citizenry than of the occupying forces. For example, the Greek Cypriot terrorist organization, EOKA killed more Cypriots—as "traitors" or "collaborators"—than it did British security forces or government officials. [17] During the early days of the Cypriot insurgency, the population, "easy-going by nature and tradition," seemed not to be taking the rebellion seriously. EOKA immediately began an offensive to punish "collaborators"; the slogan "Death to Traitors" was scrawled on walls to make the threat visible, and selected murders of Greek Cypriots "drove the point home." A concrete example of this occurred on October 28, 1955, when Archbishop Makarios publicly called for the resignation of all Greek village headmen.

> The date was symbolically chosen, for October 28th, known as Okhi ("No") Day, is the anniversary of the Greek refusal to submit to the Italian ultimatum in 1940. Only about a fifth of the headmen had responded by the end of the year, however. EOKA then went into action by murdering three headmen. Within three weeks, resignations reached 80 percent. [18]

Luis Taruc, the leader of the Huks in the Philippines, bragged that by using "old women in the town markets, young boys tending *carabaos* in the fields, and small merchants traveling between towns, every traitorous act, every puppet crime, every betrayal through collaboration" was known to the Huks. After a warning of impending "punishment," the Huks would blacklist offenders and agents were authorized to arrest or liquidate them. [19]

The Polish underground also utilized techniques of "terror propaganda" against "collaborators." Initially, specific collaborators were morally condemned in the underground press; next, they were blacklisted and perhaps "death sentences" were published. Further, the underground press frequently lumped the crimes of collaborators with the names of Nazi officials, listing the names of German officers who would be brought before war crimes tribunals after the war. In most cases, individuals thus condemned were later executed by the underground. [20]

Terror can be used in retaliation against or as a counterbalance to terror. The "Red Hand" terrorist organization used "counterterror" against the nationalists in Algeria and against Algerian agents outside the country seeking sources of financial and weapons support for the nationalist movement. The organization was composed of *colons*—Europeans in North Africa—and allegedly operated with the tacit consent of the French Government. [21]

To bring about the failure of government countermeasures, the security of the clandestine organization must be maintained. There are two ways in which an underground movement can protect its own security: "It can police the loyalty of its members and take steps to see to it that a complete picture of the movement is held only by a limited few, or it can employ the threat of terror against informers." [22]

Terror is used by insurgent organizations against their own membership in order to protect the security of their operations. It is always implicit, and often made explicit, that they who defect or betray the cause will be severely punished. To demonstrate the "reality" of the threat, undergrounds characteristically have organized "terror squads" which enforce threats and punish "traitors." Most frequently, punishment means death.

Terrorist Oaths

In Kenya in 1952, the Kikuyu tribesman who was being admitted to the select terrorist cadre of the Mau Mau had to swear himself to a "brotherhood of murder."

1. If I am called upon to do so, with four others, I will kill a European.
2. If I am called upon to do so, I will kill a Kikuyu who is against the Mau Mau, even if it be my mother or my father or brother or sister or wife or child.
3. If I am called upon to do so, I will help to dispose of the body of the murdered person so that it may not be found.
4. I will never disobey the orders of the leaders of this society.[23]

This murder oath was often accompanied by rituals involving "bestial and degrading practices," the object of which was to make the initiates become "outcasts who shrank at nothing. The acts performed were intended to be so depraved that, by comparison, the mere disemboweling of pregnant women, for instance, would seem mild." Further, this degradation would alienate the initiate from the Kikuyu community, insuring that he could never fully return to normal life and betray Mau Mau secrecy. But where such "tribal superstitions . . . proved inadequate, the gap was filled by fear of personal violence or death." [24]

Enforcing Squads

An example of underground enforcing squads is the "Traitor Elimination Corps," which was one of the first units organized by the Communist underground in Malaya. By its very existence, recruits to the underground were quickly impressed with the importance of "discipline" and with "the severity of the Party's means of enforcing discipline." However, one source cites evidence that the Malayan Communist Party also utilized lesser threats, hoping to avoid liquidating agents who broke discipline, yet always reserving the implied "escalation" of the threat of death.[25]

Interviews with numbers of surrendered Malayan insurgents revealed that although 80 percent said they feared bodily harm, the "real" reasons for their fear were "lower," more subtle threats. Nearly all said they feared most "the Party's practice of disciplining its members by depriving them of their firearms; . . . they claimed this was something to be dreaded because, without their weapons, they would be defenseless." Coupled with social ostracism, physical defenselessness threatened "the very basis of the man's sense of personal security . . . As one surrendered guerrilla reported: My friends wouldn't even look at me. I didn't know what would happen to me. I couldn't sleep at night." [20]

The Huk underground in the Philippines also used terror to maintain the security of its operations, executing its threats through a special "terror force." [27] Similarly, the FLN in Algeria enforced discipline in its urban underground networks by summarily executing "traitors" when discovered. The security of the Yugoslav Communist underground was maintained by a "secret police" called the Department for the Defense of the People (O.Z.Na.). This group provided intelligence on the behavior of underground comrades, and was authorized to liquidate those who were disloyal to the partisans.[28] During the Moroccan independence insurgency, the counterterrorist "Red Hand" guarded its secrecy by eliminating defectors through mysterious accidents."[29]

PROVOCATION

Insurgent movements frequently utilize terrorism to provoke a counteraction which may be strategically useful. This tactic was employed by the OAS, the French secret army organization in Algeria, to provoke the nationalist FLN into upsetting the "cease-fire" upon which negotiations between the French Government and the FLN were based. By indiscriminate terrorist attacks on Arab civilians, "the OAS leadership evidently believed it would so exacerbate French-Algerian relations that the Algerians would be provoked into massive countermeasures, that full-scale war would be resumed, and that no settlement would be possible." [30]

A similar use of terror to create a provocative situation is the deliberate assassination in a riot of an innocent bystander in order to create a martyr and provoke the populace into further actions against the government.

ORGANIZATION FOR TERRORISM

A key feature of terriorism is detailed preparation. To demonstrate insurgent strength and sustain momentum, early success is essential. Targets are selected so that the terrorists are free to choose the time and place that will best insure the success of the mission.[31] Effective use of terror requires a thorough knowledge of localities, people, customs, and habits; it requires extensive and secret reconnaissance activities.

"UNORGANIZED" TERRORISM

Normally; unorganized terrorism involves unplanned acts against unselected targets. Such acts are the incidental result of more generalized attacks. The distinguishing feature of unorganized terrorism is that it is committed by individuals in large units which do not have terror as their sole function.

Undergrounds usually are careful to avoid wanton acts of terrorism against the populace. Such notable experts as China's Mao Tse-tung and North Vietnam's General Vo Nguyen Giap counsel extreme restraint, advising that great care be taken to avoid bringing undue suffering to the populace and unduly alienating public opinion.[32]

SUPPORT TERRORISM

As noted earlier, terror squads are frequently used as an enforcing arm for underground political units. For this purpose, terroristic acts are specifically designed to support the underground's political goals and are usually carried out by specially trained and organized squads.

An example of this is the organization of the Communist Party underground in Malaya. Just as propaganda and political units were attached to the party's Liberation Army to assist in its "military" activities, terrorist squads were attached to the party's political arm, the Min Yuen, to enforce support of its political activities.[33] Somewhat similar to the "Blood and Steel Corps," these terrorist units consisted mostly of "trusted party thugs who, in addition to perpetrating acts of extortion and intimidation against those designated by the Party, were ordered to strengthen the treasury by engaging in payroll robberies and raids on business establishments."[34]

Another dimension of terroristic activity used in support of insurgent political goals is seen in the plan for urban insurgency adopted in 1961 by the Castro-Communist supported insurgents (FALN) in Venezuela. Here the insurgents attempted to organize terrorist units or shock brigades to serve as the catalytic agent for urban revolution. The tactic was to induce a state of paralysis and alarm within the urban public through an extended period of urban violence, eventually undermining the support and power of the government, and leading to a rapid victory. Terror squads, called Tactical Combat Units (TCU), were used for robberies, sabotage, arson, murders, and the creation of street violence and riots. The units were usually organized in detachments of about 30: 5 to 8 men engaged in the bolder terrorist actions while the rest filled lesser supporting roles.[35]

The success of many of the units' terrorist actions was largely attributable to careful advance preparation, including written operations plans; although much of the street violence appeared to be spontaneous, careful examination reveals a pattern: attacks on Venezuelan-owned properties were usually limited to robbery, whereas those on U.S. properties involved some use of incendiaries or explosives. The terrorist units operated in the mobile hit-and-run style, usually proceeding to and from their targets in stolen automobiles. Although the TCU attempted on numerous occasions in 1963 to induce an atmosphere of mass terrorism in Caracas through stepped-up sniper fire and associated acts of violence, these operations failed. Although they interrupted normal patterns of urban life, they did not succeed in producing mass terrorism of the sort that would immobilize or cripple the functioning of government.[36]

Nearly every underground movement organizes specialized terror units to conduct "professional terrorism." Such units are characterized by well-planned operations carried out by a small, highly trained professional elite, usually organized on a cellular basis. The targets are usually selected individuals of special importance, and the weapons are frequently unique and tailored to secure the safe escape of the terrorists.

In Cyprus, one of the first steps taken by George Grivas, the leader of the Greek Cypriot insurgents, was to organize a cadre of specialized terrorists. Never numbering more than 50, this small group "terrorized half a million people."[37] Significantly, Grivas developed his cadre of terrorists only from the very young. His study of Communist tactics used during the Greek insurgency convinced him that only youths in their late teens or early twenties can be molded into assassins "who will kill on order, and without question." Young men combine youthful daring and, after indoctrination, fanatical conviction, and can be made to believe they are behaving in an heroic way. Such motivation makes for an absolutely trustworthy cadre of terrorists. Grivas groomed the youths for their role as specialized terrorists through a process of escalating acts of lawlessness: first they smeared slogans on walls, then they advanced to throwing bombs into open windows or bars. Only after an extended period of testing and training were the youths given their first "professional" assignment of killing a selected target.[38]

The salient features of professional terrorism are also demonstrated in the revealing accounts of Nikolai Khokhlov, a captain of Soviet Intelligence who defected to the West in 1954.[39] Prior to his defection Khokhlov was head of an elaborate terrorist plot by Soviet Intelligence to liquidate Georgi S. Okolovich, leader of a small *émigré* anti-Soviet underground headquartered in Frankfurt, West Germany.

Dubbed "Operation Rhine," Khokhlov's two-man assassination squad was the second organized attempt by Soviet Intelligence to eliminate Okolovich. The first, in 1951, involved a group of three German Communist agents equipped with ampules of morphine, syringes, fifteen thousand German marks, and an order "to stupefy Okolovich with a morphine injection and take him into the Soviet zone. In case it was impossible to kidnap him, it was permitted to end the mission with a 'liquidation.' "[40] Because of poor intelligence, however, the mission failed: Okolovich was not to be found. Further, to add insult to injury, two of the German agents voluntarily surrendered. In planning 1953's "Operation Rhine," Soviet Intelligence placed Khokhlov, one of their own senior officers, in charge. From reports of Communist agents in West Germany, Moscow obtained pictures of Okolovich and his house, had a complete layout of his neighborhood, and detailed data about his organization. However, it was necessary for Khokhlov's team to get the "on-the-spot" intelligence, such as determining Okolovich's daily routine, how well guarded he was, where the best escape routes were, and where the assassination might best be committed.

To plan the action, the squad was brought to Moscow. Khokhlov's two

German assistants, Franz and Felix, underwent special training, learning judo and how to handle various special weapons; mastering the art of following other cars and of overtaking a target car and shooting through the windows; and learning techniques of surveillance and counter-surveillance.[41]

Khokhlov reports that they even rehearsed possible versions of the murder: a "morning" version when Okolovich was leaving his residence for work and an "evening" plan when he was returning home. In this plan,

> . . . the instructor impersonated Okolovich, I his escort. Franz practiced walking past and shooting "Okolovich" in the back after I drove away, while Felix left his car, with motor running, across the street, strolled past in time for Franz to slip him the cigarette-case weapon—and also to check on the effectiveness of Franz's noiseless shots. In case Franz missed, Felix was to use his own two-shot cigarette case; then he was to stroll back to his car, drive off leisurely, and pick up Franz a block farther on.[42]

To help insure the success of the mission, Soviet Intelligence developed special weapons: cigarette cases were refitted to encase steel blocks in which chambers for noiseless charges were drilled out. With a slight press on a small button, a chamber noiselessly discharged a bullet. The principle behind it, according to Khokhlov, was rather simple. "A steel chamber, in it, a disk. The gunpowder explodes and pushes the disk. The disk throws out the bullet and at the same moment closes the opening. All the sound remains inside."[43]

Besides giving the agents noiseless and camouflaged weapons, special consideration was given to various escape routes and plans, large sums of money were on deposit and available in case of emergency, and a variety of "cover" documents was prepared, including "authentic" Swiss and Austrian passports.

"Operation Rhine" was never executed, however. Soviet Intelligence again failed, not because of inadequate planning, but because of the human element: Khokhlov defected to the West and disclosed the entire plot.

The conduct of "Operation Rhine" nonetheless demonstrates the four essential steps in the organization of professional terrorism: intelligence, planning of action, devices, and escape-and-evasion.

Intelligence

Careful intelligence work is a prerequisite for all terrorism, particularly professional terrorism. The target is special and therefore the underground must find concrete answers to "who," "where," and "when." Intelligence must both identify the target and document his modus operandi. A local person may identify the target for outside agents who then execute the plan; while a local person can facilitate identification, it is difficult for him to commit the assassination because of local loyalties and the difficulty of escaping. Another approach is to have outside agents undertake both the intelligence work and the execution.

The difficulties of requiring local agents to assassinate local people are illustrated in an experience of the Red Hand counterterrorists in Morocco. In 1954 the underground's entire security was severely jeopardized after it assigned a local terror squad to murder the editor of the local Moroccan nationalist newspaper, *Maroc-Presse*. One of the members of the Red Hand squad was a close friend of the editor and, learning of the plot, he became remorseful and defected. The result was that the *Maroc-Presse*, "the only important French-language paper in Morocco not in sympathy with the counterterrorist Red Hand, became the best informed about their activities."[44] However, the repentant Red Hand agent soon met with sudden accidental death.

Planning

The second step in organizing for professional terrorism is planning the course of action—where and when it is to occur, and how to arrange for the target to be there. A variety of techniques has been evolved to aid the assassin in making contact with his target. If the target's schedule is totally unpredictable, sometimes a meeting which he must or will attend is called.

Devices

Another important element in organizing professional terrorism is the choice of devices to be used. Most often they are carefully tailored to the kind of job to be done—blowing up a car in which the target is riding, or assassinating the target at a public concert. Further, the devices must be designed to facilitate the terrorist's escape.

To this end, professional terrorism has developed a devilish array of weapons. Besides the cigarette-case guns described in "Operation Rhine," Khokhlov reports that Soviet Intelligence developed guns encased in fountain pens. Another popular device is the auto boobytrap technique employed by the Red Hand in Morocco. Because the technique is quick and avoids visual tampering with the car, it is considered more advanced than attaching bombs to an automobile's starter. Two magnetic holders are attached to a metal case containing a powerful charge of T.N.T. (about 280 grams) mixed with some 350 steel pellets. It is easily attached directly under the driver's seat. The detonator, a lead weight connected by a nylon thread, is placed on the exhaust pipe of the automobile in such a way that the thread is stretched taut. The moment the driver starts the motor, the vibration of the exhaust pipe topples the lead weight, which pulls on the thread and sets off the explosion. The container (wider on the top than at bottom, with a lid thinner than side walls) is designed so that the main thrust of the explosion is directed upwards at the driver's seat.[45]

Escape and Evasion

The final organizational consideration is the method of escape-and-evasion. Because professional terrorists are usually highly trained and skilled in their craft, they are obviously valuable to any underground. The problem of safe escape after executing a mission receives considerable attention.

The Communist-led underground in Greece in 1945 developed a cell of three agents whose identities were secret and who were unknown to each other. This unit, called a *synergeia*, was formed any time a professional terrorist job was required. "When the mission was accomplished, the members of the group dispersed, changed their addresses, habits, and clothes, and concocted alibis."

In Cyprus, the EOKA adopted a number of techniques to secure the escape of its terrorists. One method used was for a terrorist, posing as a journalist-photographer, to be followed down a street by two or three young girls. When he sighted the man he was to kill, he shot him in the back and immediately threw the revolver to one of the girls trailing him, who slipped it into her purse and vanished. The terrorist remained briefly on the scene under the pretext that he was a journalist.[46]

BEHAVIOR UNDER THREAT

GENERAL BEHAVIOR PATTERNS

The effect of terrorism upon individuals differs widely. Similar threats or acts affect individuals differently. Behavior patterns are affected to a large extent by personality and previously established behavior habits. For instance, those individuals who feel unable to manage or control their everyday personal affairs generally are least efficient in devising ways of meeting threat emergencies.[47]

Since personality variables affect an individual's perception of threat, the vagaries of human perception are the keys to understanding behavior under stress. It is not the "objective" character of the threat that determines an individual's behavior so much as his "subjective" evaluation of the situation.[48]

Human response to threat also varies according to the nature of the threatening situation—whether it is specific or uncertain. The terrorist may wish to have the threatened party do a particular act and may issue a highly specific threat. Where threat is clearly defined and specifically communicated to an individual, with demands, alternatives, and consequences apparent and persuasively stated, an individual's reaction is probably based upon a relatively clear assessment of known variables and he may comply out of fear of having the threat carried out.

However, the terrorist may seek to cause disruptive behavior or panic by issuing an uncertain generalized threat. The very ambiguity of the situation makes rational decision-making and assessment functions break down and leads to hysteria and panic.[49]

Some writers emphasize an important distinction between "anxiety" responses and "fear" responses: "fear is apt to produce a prompt reaction either to remove the object of fear from oneself or oneself from the object of fear," whereas anxiety "is chronic and vague . . . one does not know quite what is the cause of his anxiety and, partly for that reason, he does not know quite what to do." Thus, "the more specific the threat, the more fear-inducing it is; the more vague the threat, the more anxiety-inducing it is"— making an individual hypersensitive to ordinarily neutral situations and causing disruptive behavior.[50]

It has also been postulated that the relative intensity of threat—regardless of whether it is vague or specific—determines if a person will be able to take effective action.[51] Thus, unlike the previous "specific vs. uncertain" threat theory, where individuals respond rationally and positively to specific threat, and rather hysterically to uncertain threat, this theory suggests that whenever the magnitude of threat is great, it tends to produce an ineffective or irrational response regardless of vagueness or specificity of content. Others argue that threats or threatening acts need not necessarily grow in magnitude for terror to "heighten" or intensify"; the mere continuance of threat over a period of time is sufficient to intensify the reaction.[52]

Regardless of wheather a threat is specific or uncertain, or different in magnitude and danger, an individual's vigilance response generally evolves in five phases:[53]

1. *Recognition:* The threat, or threatening situation, is perceived by some cue or message.

2. *Probability:* An estimation of the probability of the threatening event occurring is made; the validity of threat is checked.

3. *Assessment:* The qualitative nature of the threat is assessed (whether physical pain, loss of loved ones, loss of property, etc.). There is some attempt to define the situation: nature, timing, and magnitude of threat is assessed and an estimation made of the means of coping with it, the probability of success, and the cost to the individual.

4. *Defense:* Commitment to an avenue of escape and adaptation; progressive use of a "lattice of defense," in which the failure of one defense leads to another more extensive defense.

5. *Reassessment:* When first defense routes fail, other avenues are attempted, but under the stressful or threatening situation, the individual tends to become overcompensatory, excessively sensitive, or exhibit other nonadaptive responses; psychological immobilization or breakdown may occur at this stage.

This sequence can be terminated at almost any point, and the phase "may telescope so that they are virtually simultaneous." [54] Where threat demands and consequences are apparent, rational assessment is relatively easy. Where they are ambiguous or uncertain, the likelihood of irrationality and hysteria is increased. The most debilitating factors in human response to threat are uncertainty and ambiguity, since the individual tries to resolve the uncertainty before he takes action to escape the threat. The more difficult of resolution the uncertainty appears, the more unnerved the individual becomes.

When uncertain threat leads to a state of hysteria, the individual attempts to remove the ambiguity of the threatening situation by identifying some certain source—even if the "source" has little or nothing to do with the real origin of the threat.

A corollary human response to hysteria is the predilection to suggestion —"wish-fulfillment beliefs." [55] In trying to identify the source of threat and redefine the uncertain situation, an individual succumbs to "pipedream" rumors and suggestions which explain, report, or predict some favorable outcome of the uncertain condition.

A further feature of vigilance under threat is that individuals narrow or restrict their span of attention. Becoming hypervigilant, they focus their attention on the threat and the threatener, to the virtual exclusion of other stimuli. Thus, hypervigilance leads an individual to concentrate on the demands and suggestions of the underground threatener and reduces his attention to communications from the government or security forces.

"Stress over which one has some influence can be borne with much less evidence of stress reaction . . ." If an individual can perceive no avenue of escape from a threat, he develops a sense of helplessness and this sense increases his stress reaction.[56] If the purpose of a threat is to achieve compliance with certain demands, a threat that leaves the individual with no influence over the outcome may backfire. The individual either breaks down and is unable to comply or he pursues an opposite, hostile course. For example, the Nazi policy of threatening reprisals in occupied Greece during World War II tended to operate against the German objectives of population control. Indiscriminate reprisals against the Greek populace left the individual citizen helpless to influence the outcome: guerrilla band activity near a village, over which the villager had no control, brought the threat of death. "The wanton nature of the retaliation—the picking of victims at random—meant that pro-German Greeks or their relatives suffered as much as anti-German Greeks. Under these circumstances there was little advantage in being a collaborator. As the reprisals continued they tended to give credence and prestige to the guerrillas . . ." Further, indiscriminate "burning of villages left many male inhabitants with little place to turn except to the guerrilla bands." [57]

Different behavior patterns emerge from situations in which there are conflicting threats. The individual usually succumbs to the threat which appears most imminent or is greatest in magnitude. This is illustrated by some experiences of the Philippine Army during the Huk insurgency. To counteract Huk terrorism or to dissuade a village from giving strong support to the Huk movement, the Philippine Army gathered the villagers, including the mayor and village policemen, in an open area. Approximately 200 yards away Philippine troops, in full uniform, would line up a number of "captured Huks." Then they ushered out each Huk blindfolded and executed him by bayonet. As one Philippine officer reported, "While we were killing them, some were shouting out the name of the mayor, the names of the policemen, and . . . the names of their principal suppliers. Seeing the Huks killed before their eyes, hearing themselves named as the supporters of those we had just massacred, these civilians naturally expected to be next on the death lists." [58]

In reality, the villagers had witnessed a mock execution of regular Philippine troops equipped with chicken blood and stage presence. But the "executions" had the desired effect of making the government counterthreat apparent. Afterwards, officers talked individually with the villagers, explaining that they now knew everything about the village and that those who "confessed" or cooperated would not be treated liked the captured Huks. To protect the villagers from further Huk threats, the officers established several meeting places that evening where individuals could report to give information. The threat of the government was thus made more pressing and real than the Huk terror; effective responses were obtained.[59]

Cultural factors are also a significant variable in human behavior under threat. Unique cultural mores and beliefs frequently affect an individual's sense of threat—his "state of mind" or terror. One needs only to think of the role voodoo terror plays in certain areas, such as Haiti, where the threat of the pin-in-the-doll is reputedly used with some effectiveness by agents of Haitian dictator Dr. Francois Duvalier. In Angola, it is believed that a mutilated body cannot enjoy an afterlife. The Angolian administration capitalized on this fear during the 1961 rebellion. While the tribesmen "will occasionally charge fearlessly into a barrage of machinegun fire," reports one writer, "they will think twice about attacking anyone armed with a machete." [60]

PHRASING A THREAT

On the basis of these general patterns of human behavior under stress, two essential principles of phrasing a threat can be postulated. First, there are specific threats in which the demands and consequences are communicated so that they cannot be misunderstood. Use of specific threat rests on the basic assumption that an individual confronted with persuasively stated, clear-cut demands and imminent harmful consequences, will take the line of least resistance, which is compliance. Rather than endanger himself, his family, or his property, the individual will accept, albeit reluctantly and as evasively as possible, the threatener's alternative. This, at any rate, is the theory on which the threatener bases his hopes for success.

In specific threats, the threatener seeks to secure compliance without actually being required to execute the threat. Persuasive communication, leaving little room for misunderstandings, is essential for effectiveness. For example, Viet Cong guerrillas attacking a South Vietnamese fortification during the night call over loudspeakers saying: "We only want to kill the Americans. All the rest can go free if they leave their weapons." Surrounded by superior Viet Cong forces and offered a clear alternative to further resistance, South Vietnamese militia have been known to throw down their weapons and leave the Americans to fend for themselves.[61] Occasionally the Viet Cong varies this kind of threat by distributing leaflets saying they will fire on government troops only if they are accompanied by U.S. military advisers.[62]

The tactic of "escalating" warnings has often been employed by terrorists. A mild first warning is followed by more harshly stated threats and then by an imminent show of action. World War II resistance movements in occupied Europe provide numerous illustrations of specific threats issued to "collaborators" to make them desist from supporting the Nazis. The threat often began with a "blacklisting" of the individual's name in an underground newspaper, escalated to a warning note delivered directly to him, and ended with an "execution order." All the while, the demand behind the threat was specifically stated.

If a threat is to affect a large number of people, it must be related to a clearly discernible act, so that all learn the lesson demonstrated by the enforcement of the threat. The Viet Cong terrorists in Vietnam, after several implicit or explicit warnings to a prominent villager or government official, send a signed "death sentence." When the threat—usually assassination—is carried out, "the tale of Communist omnipotence is then spread by the terror-stricken widow and children," who still have the written death sentence. "The bandwagon effect sought in these cases is to convince the inhabitants of the village that they had better obey the Viet Cong or the same fate may be theirs." [63]

Another requirement for the effectiveness of specific threat is that the threatener must have visible means of administering punishment if he is to be persuasive. The threatener must be able to determine that the threatened person did not in fact comply with the demands before inflicting punishment; otherwise the situation of a specific threat changes to one of general terror.[64]

The phrasing and structuring of specific threats vary according to the kind of compliance that is sought. Essentially, there are three kinds of threat demands. First, a threatener may choose to demand actions toward which the populace is already predisposed. This is the easiest kind—the one for which compliance is most easily secured. It may be of considerable advantage for an underground to use this tactic, for its easy success gives the threatener a disproportionate amount of credit for power and influence. For example, in Algeria the OAS, in order to demonstrate its displeasure against the French Government, demanded that all Algerians stay off the streets during the evening hours and turn off their lights, and threatened to punish anyone found on the streets during the evening hours. These demands on the populace were consonant with what an individual might do on his own during any kind of disorder.[65] Thus compliance was easy.

A second kind of demand seeks to induce an individual or group to change specific behavior by demanding alternative actions. This is more difficult than the first and requires clearly stated alternatives and persuasively stated (and perhaps demonstrated) consequences for refusing.

The third and most difficult kind of demand is one that orders an individual to refrain from a course of action he is already pursuing: the demand sharply conflicts with current behavior.

Generalized or uncertain threat is the second pattern of phrasing. The generalized threat does not delineate behavior or specify demands and consequences; these are left to the imagination of the threatened individual. Uncertain threats are used to create terror among the populace, making them vigilant and sensitive to terrorist suggestions. The threatener captures attention at a point when persons under stress are desperately searching to eliminate uncertainty and ambiguity. He may suggest escape routes and alternatives, and make compliance demands which are readily accepted in order to eliminate the uncertainty of the threat and reduce terror. Occasionally, terrorists do not even seek compliance to specific demands, but rather hope to cause "flight" or psychological and morale breakdown of a population.

The South Vietnamese Army utilized uncertain threat in a counterterror campaign, dubbed "Operation Black Eye," against the Viet Cong. Selected Vietnamese troops were organized into terror squads and assigned the task of working with rural agents in penetrating Viet Cong-held areas. Within a short time Viet Cong leaders—key members of the clandestine infrastructure—began to die mysteriously and violently in their beds. On each of the bodies was a piece of paper printed with a grotesque human eye. The appearance of "the eye" soon represented a serious threat. The paper eyes, 50,000 copies of which were printed by the U.S. Information Service in Saigon, turned up not only on corpses but as warnings on the doors of houses suspected of occasionally harboring Viet Cong agents. The eyes came to mean that "big brother is watching you." The mere presence of "the eye" induced members of the Viet Cong to sleep anywhere but in their own beds. It was an eerie, uncertain threat.[66]

Generalized terror seems to have limited effectiveness over a period of time. Uncertain threat reaches a point of diminishing returns when the populace finally either "breaks" under the stress of ambiguity or focuses hostility on some objective it perceives, correctly or not, as the source of threat.[67] Once such a "hostile belief" develops, the populace's openness to suggestion ends.

The essential distinction, then, between "specific" and "uncertain" threat is the difference between threat used to secure specifically stated demands—known and planned in advance—and threat designed to debilitate and/or sensitize a populace to later suggestions.

FOOTNOTES

[1] Colonel de Rocquigny, "Urban Terrorism," *Military Review*, XXXVIII (February 1959), p. 94.

[2] Stepniak, *Underground Russia* (New York: Charles Scribner's Sons, 1892), p. 257.

[3] Maj. Gen. Edward G. Lansdale, "Vietnam: Do We Understand Revolution?," *Foreign Affairs*, XLIII, No. 1 (October 1964), p. 81.

[4] Bernard Fall, *The Two Vietnams: A Political and Military Analysis* (New York: Praeger, 1963), p. 360.

[5] *Ibid.*, pp. 360–61.

[6] Denis Warner, *The Last Confucian* (New York: Macmillan, 1963), p. 137.

[7] Robert B. Rigg, "Catalog of Viet Cong Violence," *Military Review*, XLIV (December 1962), p. 25.

[8] U.S., Department of State, *A Threat to the Peace: North Vietnam's Effort to Conquer South Vietnam*, Part II (Washington, D.C.: U.S. Government Printing Office, 1961), p. 101.

[9] Col. Ismael Lapus, "The Communist Huk Enemy," *Counter-Guerrilla Operations in the Philippines, 1946–1953* (Seminar) (Fort Bragg, N.C.: June 15, 1961), p. 18.

[10] Rigg, "Catalog of Viet Cong Violence," p. 25; and Roger Trinquier, *Modern Warfare* (New York: Praeger, 1964), pp. 16–17.

[11] Wesley R. Fishel, "Communist Terror in South Vietnam," *The New Leader*, III, Nos. 27–28 (July 4–11, 1960), p. 14.

[12] Brian Crozier, *The Rebels* (London: Chatto and Windus, 1960), p. 172.

[13] Paul Jureidini, *Case Studies in Insurgency and Revolutionary Warfare: Algeria 1954–62* (Washington, D.C.: Special Operations Research Office, 1963), pp. 98–99.

[14] Lucian W. Pye, *Guerrilla Communism in Malaya* (Princeton, N.J.: Princeton University Press, 1956), p. 104.

[15] Crozier, *The Rebels*, p. 176.

[16] *Ibid.*

[17] *Ibid.*, p. 170.

[18] *Ibid.*

[19] Luis Taruc, *Born of the People* (New York: International Publishers, 1953), pp. 134 and 137.

[20] Interview with former member of the Polish underground, Washington, D.C., April 27, 1965.

[21] Joestan Joachim, *The Red Hand* (London: Abelard Schuman, 1962), pp. 47–49.

[22] Gershon Rivlin, "Some Aspects of Clandestine Arms Production and Arms Smuggling," *Inspection for Disarmament*, ed. Seymour Melman (New York: Columbia University Press, 1958), p. 192.

[23] Crozier, *The Rebels*, pp. 169–70.

[24] *Ibid.*, p. 177.

[25] Pye, *Guerrilla Communism in Malaya*, p. 252.

[26] *Ibid.*, pp. 252–53.

[27] Taruc, *Born of the People*, p. 134.

[28] See U.S., Senate, Committee of the Judiciary, *Yugoslav Communism—A Critical Study* (Washington, D.C.: Government Printing Office, 1961), p. 124.

[29] Joachim, *The Red Hand*, pp. 47–49.

[30] James Eliot Cross, *Conflict in the Shadows: The Nature and Politics of Guerrilla War* (Garden City, N.Y.: Doubleday and Co., 1963), p. 54.

[31] Colonel de Rocquigny, "Urban Terrorism," p. 95.

[32] See Vo Nguyen Giap, *People's War, People's Army* (Washington, D.C.: Government Printing Office, 1962), pp. 65ff.

[33] Fred H. Barton, *Salient Operational Aspects of Paramilitary Warfare in Three Asian Areas*, ORO–T–228 (Chevy Chase, Md.: Operations Research Office, 1953), p. 40.

[34] Pye, *Guerrilla Communism in Malaya*, p. 88.

[35] Atlantic Research Corporation, *Castro-Communist Insurgency in Venezuela* (Alexandria, Va.: Georgetown Research Project, Atlantic Research Corporation, 1964), pp. 21–25.

[36] *Ibid.*, p. viii.

[37] Dudley Barker, *Grivas: Portrait of a Terrorist* (New York: Harcourt, Brace and Co., 1959), p. 142.

[38] *Ibid.*, p. 80.

[39] Nikolai Khokhlov, *In the Name of Conscience* (New York: David McKay, 1959), p. 365.

[40] *Ibid.*, p. 194.

[41] *Ibid.*, pp. 229–30.

[42] *Ibid.*, p. 230.

[43] *Ibid.*, p. 232.

[44] Joachim, *The Red Hand*, p. 48.

[45] *Ibid.*, p. 200.

[46] Barker, *Grivas*, p. 145.

[47] Stephen B. Withey, "Reaction to Uncertain Threat," *Man and Society in Disaster*, eds. George W. Baker and Dwight W. Chapman (New York: Basic Books, 1962), p. 118.

[48] *Ibid.*

[49] Neil J. Smelser, *Theory of Collective Behavior* (New York: The Free Press of Glencoe, 1963), pp. 83ff.

[50] James C. Davies, *Human Nature in Politics* (New York: John Wiley and Sons, 1963), pp. 67–68.

[51] *Ibid.*, p. 68.

[52] Withey, "Reaction to Uncertain Threat," p. 121.

[53] *Ibid.*, pp. 95ff.

[54] *Ibid.*, pp. 105–106.

[55] Smelser, *Theory of Collective Behavior*, p. 84.

[56] Withey, "Reaction to Uncertain Threat," pp. 95 and 121.

[57] D. M. Condit, *Case Study in Guerrilla War: Greece During World War II* (Washington, D.C.: Special Operations Research Office, 1961), p. 268.

[58] Medardo T. Justiniano, "Combat Intelligence," *Counter-Guerrilla Operations in the Philippines, 1946–1953*, Seminar (Fort Bragg, N.C.: June 15, 1961), pp. 47–48.

[59] *Ibid.*, p. 48.

[60] Ronald Waring, *The War in Angola—1961* (Lisbon: Silvas, 1961), p. 27.

[61] *The Evening Star* (Washington, D.C.), February 14, 1965, p. A–1.

[62] "What's News: South Vietnamese Communists," *The Wall Street Journal*, CLXIV (August 12, 1964), p. 1.

[63] Fishel, "Communist Terror in South Vietnam," p. 14.

[64] Andrew R. Molnar, *et al.*, *Undergrounds in Insurgent, Revolutionary and Resistance Warfare* (Washington, D.C.: Special Operations Research Office, 1963), p. 104.

[65] *Ibid.*, pp. 104–105.

[66] Malcolm W. Browne, *The New Face of War* (New York: Bobbs-Merrill, 1965), pp. 119–20.

[67] Smelser, *Theory of Collective Behavior*, pp. 83 and 101ff.

CHAPTER 12

SUBVERSIVE MANIPULATION OF CROWDS

The underground use of subversively manipulated crowds and civil disturbances has added a new dimension to the problem of maintaining internal security. The difference between civil disturbances which are subversively manipulated and those which are not can be expressed in terms of objectives. Strikes, riots, and demonstrations usually have limited goals, such as better working conditions or social changes. The aims of the underground movement are the overthrow of the government and the seizure of power.

The manipulation of crowds and civil disturbances is just one of the means used to accomplish the objective of seizing power. The internal security forces, who bear a major share of the burden of maintaining order, should understand that the control of subversively manipulated crowds requires special considerations. Standard priorities of force may be adequate for dispersal of ordinary civil disturbances, but in dealing with a subversively controlled riot, internal security forces must be alert to situations or acts which compel them to respond in ways that the subversives can politically exploit.

The security forces of a nation are usually composed of paramilitary and military units and civil police. The function of maintaining internal security may be performed by one or any combination of these forces.

STRATEGY AND TACTICS

One method for improving our understanding of this phenomenon is to study the strategy and tactics of subversion. Underground strategy is to separate the existing government from its base of power by capturing the institutional supports upon which it rests, by alienating mass support from the government, and by overtaxing internal security forces with problems of unrest. It has been said that "revolutionists in modern society do not so much 'seize' power as destroy and re-create it." [1] The simple creation of disorder does not automatically bring an underground group to power. It can, however, create a vacuum into which new organizational instruments of power can move.

When the habit of obedience to law breaks down among a populace, a tense, highly emotional state ensues, which gives the underground a chance to channel dissatisfactions. The tactics of internal subversion involve the subgoals, methods, and techniques of creating social disorganization. One such tactic is the creation or manipulation of crowds and civil disturbances for the purpose of advancing the overall strategy.

The population target group of the underground may be a large minority with certain crystallized grievances, or it may be a highly organized group, such as a labor union or regional political party, that is continually bargaining for a more favorable position in society.

The community conflicts fomented by or capitalized upon by the underground follow a pattern. One of the peculiarities of social controversy is that it sets in motion its own dynamics which carry it forward on a path which may bear little relation to its beginnings.[2] When there are deep cleavages of values or interests in a community, specific grievances have a tendency to give way to general issues, and groups that set out to protest a specific issue end up disagreeing with the entire civil administration.

As a controversy develops, new and different issues, unrelated to the original, frequently emerge or are introduced. These added elements may reflect deep-seated prejudices or individual grievances. They are characteristically one-sided, so that response can be in only one direction, and are structured so as to capture the attention of the members of the community. The new issues increase solidarity among old members and attract new members.

Another significant complication is the progression from simple disagreement to violent antagonism. The generation and focusing of hostility can sustain social conflict without the aid of specific facts and issues. This results in a tendency to see the opposing group as all bad and the interest group as all good.

An interesting change takes place in the social organization of the community. Long-standing relationships among individuals and groups are terminated and new groups polarize around the issues in conflict. New leaders tend to take over the dispute. The social geography of the community is altered. At this point in the social process the underground attempts to manipulate at least one of these groups by directing and channeling its grievances, grasping the leadership, and speaking for it.

RIOTS AND DEMONSTRATIONS

The subversive manipulation of crowds and civil disturbances involves a relatively small number of undergrounders who try to guide and direct "legitimate" protests. They attempt to direct the crowd toward emotional issues and arouse them against authority.

The emotional perceptions and beliefs of the crowds that participate in civil disturbances often do not coincide with objective reality, and the individuals involved do not realize that their grievances are being manipulated in politically subversive ways. The following accounts of riots and demonstrations illustrate underground attempts to incite or exploit civil disturbances.

COLOMBIA (1948)

The incident called El Bogotazo (Blow at Bogotá) demonstrated the effectiveness of the riot technique. It is particularly significant in that the techniques for provoking street crowds into rioting learned in the demonstration are used as training material for revolutionaries. A Venezuelan Communist defector from Castro's training school—the Tarara Training Center in Cuba—has reported that the classroom materials include diagrams of the tactics and descriptions of how the crowds were manipulated in the 3-day riot that wrecked the Ninth Inter-American Conference and left the Colombian capital in ruins.[3]

The Ninth Inter-American Conference had been called for March 30, 1948, to discuss the pledge among member nations for mutual defense and resistance to the threat of international communism. According to testimony before the Judiciary Committee of the 86th U.S. Congress, intercepted Communist communiques reveal that the party immediately went into action; the combined forces of the Latin American Communist Party apparatus set about making plans to disrupt the conference.[4] A high-ranking official of the Colombian Communist Party (which claimed at that time a membership of 10,000 out of a total population of 11 million) said that the Inter-American Conference must be blocked, but that this action was not to be known as a Communist activity; he admonished the party to refrain from open activity so as not to jeopardize or curtail party functions.

The Communists devote considerable effort to preparing for any proposed riot; they seldom rely exclusively on spontaneity or accidental occurrences, even though they attempt to capitalize on such events. By January 29, 1948, arms and explosives had already been stored in 17 houses. A Communist dispatch dated February 2 included the information that plans called for organization of mass public meetings, organization of 16 meetings of cells in outlying districts, recruitment of new members to the party, organization of 15 syndicates and unions, further organization of cells within the syndicates, and distribution during the conference of 50,000 handbills and 3,000 posters. A committee of the Communist Party was assigned to supervise these arrangements. A dispatch dated March 30 laid out the program of agitation and attacks upon the United States, Chilean, Brazilian, and Argentine delegations, all of which were especially anti-Communist. During the middle of the first week in April, the Communist-controlled Latin American Conference (CTAL) adopted resolutions in Mexico City condemning the conference.

On April 9, a well-known figure, Dr. Jorge Gaitan, was killed by four bullets from a revolver fired by an unidentified person. Dr. Gaitan, a 47-year-old lawyer, was the leader and former presidential candidate of Colombia's liberal movement. Although it was reported that the Communist Party had supplied money through an intermediary for the support of Dr. Gaitan and his movement, he had maintained an independent attitude toward the Communists. Rumors of Communist plans to disrupt the conference caused Dr. Gaitan to publicly repudiate all acts against the conference, saying that these were acts against democracy and the unity of the Americas. Although his assassin was never identified, both the personality of Dr. Gaitan and the circumstances surrounding his death inspired at least one observer to say that the Communists, needing an appropriate victim whose death could prevent the holding of the conference, selected a prominent person.

Colombian President Ospina Perez also suggested that the man who killed Dr. Gaitan apparently had Communist affiliations and that the entire affair was a Communist maneuver.

A wave of mass violence was triggered by the assassination, which took place within the sight of thousands. The Communists channeled the high emotions into anti-U.S. feelings and acts of violence against U.S. property and individuals. Within 15 minutes of the attack on Dr. Gaitan, radio broadcasting stations in Bogotá had been taken over and the Communists were issuing instructions and inciting the people to revolt against the government, the conference, and Yankee imperialism. Orders were given to plunder arms depots, hardware stores, gunsmith shops, department stores, government buildings, police precincts, and army barracks, and to organize a "popular militia." The radio also transmitted orders to specific individuals to assault specific places and gave locations where additional weapons could be obtained. Instructions were given on how to manufacture Molotov cocktails. During the broadcasts, fighting could be heard in the background, reportedly between the Communists and a group of students, for control of the radio facilities; the armed Communists forced the students out of the station. By controlling communications, the Communists could incite attacks against the symbols and instruments of power within the government. Within each group of demonstrators in the crowds were organized agitators chanting similar slogans. Prompted by the Communist agitators, a crowd entered the parliament building where the Inter-American Conference was being held and destroyed most of the interior. The rioters concentrated on destroying offices of the Chilean and United States delegations. Mob action almost completely suspended transportation, created obstacles for the police, and made the crowd that much more difficult to control.

Led by the Communists, less than 5 percent of the population carried on the riot for 3 days. Shops, churches, public utilities, and institutions of public service were attacked. Red flags were evident throughout the crowd and in every group orders could be heard directing the mobs. The word "abajo" (down with) was heard frequently. There was heavy sniper fire.

The first burnings of buildings may have been simple, random gestures of protest, but a consideration of which buildings were burned showed a subversive pattern. On the afternoon of April 9, the Ministry of National Education, the Ministry of Justice, the Ministry of Foreign Affairs, the Palace of Justice, the Ministry of Government, the Episcopal Palace, the detective headquarters, and the Identification Section for natives and foreigners were all attacked and burned. The Confederation of Colombian Workers, which represented 109,000 organized workers, called for a general strike throughout Colombia.[5]

After declaring a state of siege and imposing martial law, President Perez eventually restored order. But the Communists had achieved their tactical objective of disrupting the conference and, in the process, had effectively demonstrated the practicability of their methods.

VENEZUELA (1960–62)

The significance of the Venezuelan Castro-Communist insurgency is its expansion of the tactic of riot into strategic prolonged urban violence, planned to paralyze the functioning of government.[6] This strategy, unlike long-term, rural-based insurgency, was aimed at achieving a "rapid victory." The urban insurgency movement first focused on organization— the establishment of secret "activist nuclei" in Caracas, around which terrorist groups could later operate.

Then, beginning in early 1961, it launched a program of violence in the form of riots and subversive demonstrations. According to the Venezuelan Ministry of Interior Relations, a total of 113 "significant riots" occurred in the course of the year. The most common technique was to mass hundreds of students, largely those enrolled in leftist youth organizations of the Central University and various secondary schools, and move them toward the center of the city. En route, buses and automobiles were burned and flaming street barricades were erected. Older students were frequently armed with Molotov cocktails, and some carried pistols or acid. Students posted within the university grounds (which are off-limits to police because of the university's autonomous status) added to the confusion and casualties with sniper fire.

The third stage of the urban insurgency featured the introduction of "shock brigades"—small terrorist units (called Tactical Combat Units) which served as auxiliaries to the main body of student rioters. While security forces were occupied with the main riot center, "shock brigades" fanned out to create numerous points of street violence with sniper fire, burning of vehicles, blocking of traffic, and destruction of property.

Although the Communist underground markedly improved its techniques of street violence through increasingly effective organization, the repeated riots and urban violence failed to achieve a rapid victory. The student riots in Caracas were of little practical value because the general public stayed aloof from the violence and was content to wait for government forces to restore order. The Communists never succeeded in employing the two most formidable techniques of urban insurrection—large-scale rioting and a revolutionary general strike—mainly because they could never coalesce public opinion behind them.

SOUTH VIETNAM (1964)

In South Vietnam, the Viet Cong have used tactics similar to those used in Bogotá in 1948. Captured documents and questioning of prisoners have disclosed several phases in developing riots.[7] In the first phase, the Communists established safe zones in sections of the city. Within these areas, they stored arms and identified places where insurgent personnel could gather secretly. The Communist infiltrators were to disguise themselves as students or workers and were then instructed to infiltrate legal and semi-legal clubs and associations or, for that matter, any organization which could be used as a vehicle for countergovernment propaganda.

In the second phase of the operation, Communist youth groups, armed with clubs and knives for "self-defense," instigated street quarrels to create tension and manufacture further incidents. Agitators were to mix with the crowd, yelling inflammatory slogans to whip up excitement. The instructions stressed the importance of creating martyrs to focus the crowd's attention on the injuries and deaths caused by the government forces. Armed groups were instructed to assassinate city officials and seize police weapons.

The Viet Cong underground identified shopkeepers and homeowners who would be willing to shelter demonstrators as they fled from the police and who would hide the cadres of agitators during police searches.

The Viet Cong clandestine radio and Radio Hanoi called for urban uprisings in daily broadcasts. During one mass demonstration a 15-year old-youth was killed. The next day, during a dramatic funeral procession for him, paratroopers seized 10 youths who had knives hidden under their shirts, thus breaking up plans for new violence.

PERU (1964)

In Peru, a Chamber of Deputies committee investigating the so-called Sicuani Massacre found that Cuban-directed Communists had planned and forced unwilling peasants to join in "land-grab leagues." [8] Despite the extreme socioeconomic underdevelopment of the Department of Cuzco and the unfortunate land tenure conditions—alone sufficient to enable the Communists to obtain support from the peasants—it was found that Communist peasant leaders threatened to inflict bodily harm on unwilling peasants unless they demonstrated for the land seizures. During the demonstrations women and children were placed in front of the men in peasant encounters with the civil guards. The investigating committee found that after stones had been thrown at the police, one of them hitting the Chief of Police, the civil guard units fired into the crowd, resulting in the "massacre."

COLOMBIA (1965)

On January 21, 1965, Colombian President Valencia made a speech announcing that government forces had reestablished control after several days of rioting in Bogotá. The riots had the appearance of a general strike but in his speech the President pointed out to the Colombian people the subversive nature of the riots and the subsequent wave of bank holdups and kidnappings. The government had intercepted Communist communications directing the course of the demonstrations. These communications revealed that Communist youth groups had been instructed to mobilize the masses through meetings in centrally located places. Factory cells had been instructed to incite strikes, but not to let the connection with the Communist Party show. The cadres had been instructed to avoid confrontation with the police and the possibility of being jailed. They had been instructed to launch a popular front demanding the abolition of the sales tax, the resignation of the government, and the installation of a popular junta with elements from all political parties. They had been further instructed to organize factory meetings at district and zonal levels and to speak for a single front made up of liberals, conservatives, and Communists. They had also been instructed to launch attacks on U.S. business firms, to release prisoners from the Bogotá jail, and to invite army personnel and armed police to join them. The targets selected for seizure were the National Radio Station, press telephones, and public water and power works. The Communists had printed posters and appealed for the formation of "public salvation committees" in all cities and villages throughout the country which were willing to confront the situation and take on the responsibilities of government.

PHASES OF SUBVERSIVE MANIPULATION

Subversively manipulated civil disturbances may be considered as having four phases: (1) the precrowd phase; (2) the crowd phase; (3) the civil disturbance phase; and (4) the post civil disturbance phase.

THE PRECROWD PHASE

In the preparation or precrowd phase, the underground elements are primarily concerned with *building an organization*. Lenin maintained that training a network of agents for the rapid and correct distribution of literature, leaflets, and proclamations accounted for the greater part of the work of preparing for a demonstration or uprising. He concluded that it is too late to start organizing literature distribution at the moment when a strike or demonstration is about to start.[9]

Selected individuals are given special *training* in the subversive manipulation of crowds. They are taught how to build barricades and conduct street fighting, how to mobilize blocks in the city and workers in plants, how to develop a local strike into a general strike and general strike into a city uprising, and how to coordinate these into a national uprising.[10] Outside specialists are often brought in to direct the training activities.[11]

Some sort of *planning* on the part of the underground must take place. It may vary from rudimentary to highly sophisticated. Underground agents are instructed to infiltrate target groups by joining formal organizations, clubs, or any association which gives them access to such audiences.[12]

Next comes the *selection of a population target*. It is chosen primarily for its potential to bring about community conflict or increase its intensity. Any group that is not susceptible to manipulation, at least after some preparation, is not considered to be a part of the "masses," in Communist terms. Groups identified by their common interests (e.g., ethnic minorities, labor, farmers) offer great potential for covert manipulation, because attention can be centered on bread-and-butter issues rather than on complicated ideological sensitivities.[13]

The desired change in attitude of the members of the target group is usually accomplished through distribution of *selected communications*, the contents of which are designed to increase anxiety and emotional stress. Word of mouth, radio, telephone, and leaflets and other printed material have been effectively used. Pistols, rifles, materials for making Molotov cocktails and explosives, and other weapons, such as clubs and lengths of pipe, and also handbills, signs, armbands, and banners, must be acquired and stored. In recent riots, such weapons as handguns, rifles, and Molotov cocktails have been employed. The rifle has proved particularly useful for creating additional chaos by the killing or wounding of members of the crowd or internal security forces from relatively safe distances.[14]

Arrangements for members of the underground group to flee the area must be completed. These consist primarily of establishing *routes of escape* containing safe houses or other hiding places. Safe zones are established with householders and shopkeepers where demonstrators may seek cover when fleeing from the police.[15]

In places where demonstrations or strikes can be planned in advance, the underground mounts a campaign directed at *preconditioning* target groups. Chosen themes are constantly repeated. By concentrating on local and specific grievances, a group is conditioned to phrases and slogans to which its members may later react under conditions of emotional stress.[16]

THE CROWD PHASE

The indispensable element in civil disturbances is the crowd: not just any crowd, but a crowd made up of individuals who have been conditioned either by subversive manipulation or by other events.

Organizations

There are several ways to assemble a crowd. Cell members infiltrate mass organizations so that strikes or mass meetings can be changed into armed demonstrations. There are built-in sanctions within labor unions or other disciplined organizations which can be used to punish members who do not comply with the decisions of the organization. Therefore, if the infiltrated union or organization calls for a strike or demonstration, its members can be brought into a particular place at a particular time. Student groups are highly volatile on many social issues and can be induced to participate in demonstrations for the sheer excitement.

Informal Gatherings

Demonstrations can be brought about at parades, street parties, dances, or during normal rush-hour periods. During Vice President Richard M. Nixon's trip to Caracas, Venezuela, on May 13, 1958, crowds of students and other onlookers were turned into rioters by slogans and chants. Eventually they attacked Nixon's car with heavy rocks, jagged cans, eggs, and tomatoes, and beat the windows with clubs. The traffic jam which permitted Vice President Nixon's car to be attacked was prearranged. Two trucks collided and the drivers then just walked away. The street was also packed with banner-carrying youngsters primed for action.[17]

Hired Demonstrators

In addition to the local people at the demonstration, against Vice President Nixon, many top Latin American Communists from nearby countries had converged on the capital. Party agents went into areas where criminal elements lived and hired as many as they could, arming them with long wooden clubs and iron bars. Gustavo Machado, a Venezuelan Communist Party leader, later admitted to organizing the demonstration in Caracas.[18]

Before James Hagerty's trip to Japan to prepare for the visit of President Eisenhower, the Communists began planning and organizing the demonstrations that were to greet him. To assure large participation in the riots, the organizers paid 1,000 yen (approximately $2.78) to persons who would attack Hagerty's car, and 350 to 500 yen to persons for general participation. In addition, workers from the Sohyo labor federation received a half-day's salary for participation. Applicants were recruited from the employment offices to such an extent that police were able to predict the large demonstration by the absence of applicants in the offices.[19]

The Precipitating Event

The precipitating event which results in the formation of a crowd depends for a great deal of its effectiveness upon communication, especially upon distortion of the event. If the event is fabricated by the agitator, the distortion is built in. If the event is factual, he makes it fit his issues or capitalizes on the natural distortion which accompanies word-of-mouth communications. The precipitating issue or event can be a martyred individual, a report of police brutality, or a symbolic act such as the desecration of a flag.

Mob Management Techniques

Although the Communist Party of Iraq in the militant period of 1948–50 was a tiny minority, it succeeded in creating the impression of large numbers and great support by successful mob management. The Communist elements were organized into an external command, well removed from the activity, which could observe the demonstration, and an internal command located in the crowd, which was responsible for directing the demonstration. There were bodyguards who surrounded and shielded the internal command from the police and, if necessary, facilitated their escape; messengers who carried orders between the internal and external commands; shock guards who were armed with clubs and acted only as reinforcements, creating diversionary violence when Communists became engaged by the police thus permitting them to escape; and banner carriers who switched from banners expressing general grievances to those reflecting direct Communist propaganda at the appropriate time. The cheering sections consisted of special demonstrators who had rehearsed the slogans and chants and the order in which they were to be raised.[20]

The Agitator

After the crowd has been formed, the agitator assumes a significant role.[21] His function has been described as bringing to flame the smouldering resentment of his listeners through emotional appeals and then giving social sanction to their actions. The agitator in the crowd plays upon the audience's suspicion of things they do not understand. He points out that there is material abundance for everyone but that the crowd does not get its proper share. He generally points to a premonition of disasters to come and plays upon the fears of individuals and the uncertainty of life in the community. The agitator then points to the politicians and the police as representatives of government and alludes to fraud, deception, and falsehoods among them. The agitator seldom invents issues, nor does he have to, since his appeals are vague and he plays upon the basic emotions of fear and insecurity. The agitator seldom justifies his facts, nor does he need to, since he chooses common emotional themes common to all men.

The Riot Leader

After the crowd has been emotionally aroused, some event must set it in motion. Often it begins its riotous activity by following a leader who merely shouts "follow me, let's go." The internal command assumes the leadership role if an emergent leader does not arise spontaneously. The event which sets the crowd in motion may, like the precipitating incident that brought them together, be either factual or fabricated.

Small groups have face-to-face communications and interaction. In large groups, however, communications come through second- and third-hand sources. This is conducive to the spirit of rumor. It has been said that "no riot ever occurs without rumors to incite, accompany, and intensify the violence." [22] Rumor tends to mobilize collective action. It may be communicated through gestures, casual conversation, or mass media. It may be triggered by an actor who sets an example by what he does and spontaneously becomes the leader of a group or it may be planned by the underground organization.[23] Precipitating events gives generalized beliefs immediate substance. Rumors as they are related to beliefs tend to restructure the ambiguous and uncertain situation and to explain it for the individual who is participating in the crowd. They help to put facts into place.[24]

THE CIVIL DISTURBANCE PHASE

Maintaining Emotional Excitement

Once the destructive action of the crowd is under way, the agitator tries to maintain the level of emotional excitement. This can be accomplished in various ways. Cheerleaders can chant rhythmic and inspiring phrases or songs. Slogans can be displayed and banners unfurled. "Booster" incidents can be created or capitalized upon—a rather universal type of booster activity is the looting of stores and shops. Bank holdups and kidnappings are also carried out during the chaos. Other acts—such as the verbal abuse and stoning of police—which permit the individual to release aggression and hostility against the symbols of authority also increase the emotional involvement. In the Panama riots in January 1964, 400 to 500 people threw stones and Molotov cocktails when assaulting the home of Judge Guthrie F. Crowe. Later, they attacked the railroad station with the same weapons. The Molotov cocktails used during the rioting must have been made specifically for the riots, said the investigating committee appointed by the International Commission of Jurists—but when, where, and by whom was not disclosed.[25]

Creation of Martyrs

The creation of a martyr has a sustaining effect upon destructive crowd activity. Subversive elements do not deplore bloodshed and violence. Attacks are made upon the internal security forces in order to provoke retaliation. If this fails, a rifle in the hands of a sniper can assure a victim. A martyr turns an ordinary grievance into an emotional crusade. Individuals who cannot easily identify with abstract issues readily empathize with the emotional demands brought on by apparent injustice to or "brutal" attacks on innocent people.

The police are provoked by insults and attacks into hostile, aggressive responses which will lead to the injury of women, children, or other "innocents" who happen to be in the crowd. This is then dramatized and used as a new emotional issue for which the crowd must seek retribution. The accidental death of the 22-year-old daughter of a university professor was a major issue in the Tokyo riots leading to the cancellation of President Eisenhower's visit to Japan in May 1960. The girl was trampled to death during a clash between a group of demonstrators and the police. The police were blamed for her death, and an elaborate funeral was staged at which thousands of demonstrators were present. Many Tokyo students and university professors took part in a large rally and in subsequent demonstrations at the Diet, the Prime Minister's residence, and the Metropolitan Police Board to express their concern and indignation over the police and government actions. The demonstration was not only instrumental in canceling President Eisenhower's trip, but probably had some impact upon the subsequent fall of the Kishi government.[26]

Counterpolice Activity

Police and army counterriot tactics are studied by the planners so that steps can be taken to circumvent them. Routes usually taken by internal security forces are blocked with barricades, overturned vehicles, and debris. Attacks upon police stations and their communications systems serve to disrupt police countermeasures. Cadres are usually guarded by strong-arm squads and avoid confrontation with the police so they will not be jailed. Appeals to army or police units not to attack their own countrymen are especially effective when there are children, women, war veterans, or students in the front ranks of the crowds.[27]

POST-CIVIL DISTURBANCE PHASE

After a civil disturbance has subsided, underground elements use a variety of means to capitalize on the situation. One way of maintaining the interest and emotional involvement of the population is a *24-hour general strike.* Workers, especially those in key industries and utilities, are encouraged to protest against the government by staying away from work 24 hours. This is time enough to interrupt vital utilities and affect the entire population. Individuals and property owners are faced with the dilemma of going about their normal routine and facing violence or staying home for a day. Factory cells incite union members to stay off the job. They seek union sponsorship or usurp authority by making unauthorized announcements or by calling individual members.

To turn specific issues into grievances against the government, the underground makes appeals to all workers to join in a *united front* against the government. Cells in various factories, districts, zones, and businesses demand that their organizations support the strike in the form of a united front.

Agitators attempt to capitalize upon the contagious effect of civil disturbances by spreading the *violence* and creating new incidents in nearby areas. Attacks upon symbols of authority, such as police stations and the offices of local officials, increase the intensity of the disorder. If possible, radio stations, newspapers, water, and power services are seized. Newspapers and radios spread the rumors, and control of water and powerplants spreads social disorganization and fear. In the Bogotá riot of 1948 and in the Bolivian riots of April 1953, appeals were made to workers, peasants, and trade unionists to form armed "people's militias" and correct the "injustices" of the government.[28]

In order to demonstrate the uncompromising postion of the government, the *demands* against it are usually vague and impossible to meet. Original issues, such as higher wages or repeal of a sales tax, are now changed to antigovernment demands. A call is made for the release of political prisoners, and the police and army are asked to join the rioters. It is customary to insist on nothing less than the complete overthrow of the existing government. These demands can be articulated in protest meetings that keep the public aroused and involved. Committees are formed in every village or city to protest government action. Every attempt is made to get notable and respected citizens to lend their names to the protest.

FOOTNOTES

[1] George S. Pettie, *The Process of Revolution* (New York: Harper and Brothers, 1938), p. 5.

[2] James S. Coleman, *Community Conflict* (Glencoe, Ill.: The Free Press, 1957), pp. 9–10.

[3] Juan DeDios Marin, "Inside A Castro 'Terror School'," *The Reader's Digest* (December 1964), pp. 119–23.

[4] U.S., Congress, Senate, Committee on the Judiciary, Subcommittee to Investigate the Administration of the Internal Security Act, *Hearings on Communist Threat to the United States through the Caribbean*, 86th Cong., 1st Sess., 1959, pp. 115–24; see also Nathaniel Weyl, *Red Star Over Cuba* (New York: Devin-Adair, 1960), pp. 1–38.

[5] Donald Marquand Dozer, "Roots of Revolution in Latin America," *Foreign Affairs* (January 1949), pp. 286–87.

[6] Peter Grose, "Viet Cong Intensifying Efforts To Spread Disorder in Saigon," *The New York Times*, December 2, 1964.

[7] *The Hispanic American Report*, Stanford University, May 1964, p. 249.

[8] Atlantic Research Corporation, *Castro-Communist Insurgency in Venezuela* (Alexandria, Va.: Georgetown Research Project, Atlantic Research Corporation, 1964), pp. 29 and 125–28.

[9] Vladimir I. Lenin, "Letter to a Comrade on Our Organizational Tasks," *Collected Works*, Vol. 6 (Moscow: Foreign Languages Publishing House, 1961), p. 240.

[10] U.S., Congress, Senate, *Communist Threat*, p. 138.

[11] Eugene H. Methvin, "Mob Violence," *Military Review*, XLII (March 1962), pp. 29–40.

[12] Grose, "Viet Cong Intensifying Efforts."

[13] Philip Selznick, *The Organizational Weapon: A Study of Bolshevik Strategy and Tactics* (Glencoe, Ill.: The Free Press, 1960), p. 83.

[14] Grose, "Viet Cong Intensifying Efforts."

[15] *Ibid.*

[16] Methvin, "Mob Violence."

[17] U.S., Congress, Senate, *Communist Threat*, pp. 126–32.

[18] Methvin, "Mob Violence."

[19] *Ibid.*

[20] *Ibid.*

[21] Leo Lowenthal and Norbert Guterman, "Portrait of the American Agitator," *Public Opinion and Propaganda*, eds. Daniel Katz, *et al.* (New York: The Dryden Press, 1954).

[22] G. W. Allport and L. Postman, *The Psychology of Rumor* (New York: Henry Holt, 1947), pp. 193–96.

[23] Neil J. Smelser, *Theory of Collective Behavior* (New York: The Free Press of Glencoe, 1963), p. 11.

[24] *Ibid.*, p. 81.

[25] International Commission of Jurists, *Report on the Events in Panama, January 9–12, 1964* (Geneva: International Commission of Jurists, 1964), p. 21.

[26] Methvin, "Mob Violence."

[27] See Rex Applegate, *Kill or Get Killed* (Harrisburg, Pa.: The Stackpole Co., 1961), pp. 375–76.

[28] U.S., Congress, Senate, *Communist Threat*, pp. 117–22.

PART V

PARAMILITARY OPERATIONS

INTRODUCTION

Many of the functions performed by the underground can properly be called military activities. The exigencies of present-day insurgencies preclude a monopoly of the performance of military functions by regular mainforce units. A member of a regional force or local militia is often a peaceful fisherman by day; only at night does he don his uniform and conduct raids. The political and military activities of an insurgency overlap both in function and in personnel.

Usually inferior to the government security forces in number and resources, the underground must use every opportunity and capitalize on every advantage. To do this requires careful planning, adequate intelligence, and effective means of escape-and-evasion. Intelligence relies on reconnaissance, underground infiltration and on the cooperation of the part-time underground in the village; planning requires the intelligence provided by the villagers and a careful analysis of the enemy forces; performance of ambushes, raids, and sabotage depends on adequate intelligence, careful planning, and skillful use of well-trained underground elements. Once the raid or ambush is over, provision must be made for withdrawal, exfiltration, and dispersal. The escape-and-evasion nets provide for the safety and survival of the underground members after the mission is over.

The following chapter will discuss the activities of planning and intelligence and describe three areas of underground activity in which their operational aspects have been applied: ambushes and raids, sabotage, and escape-and-evasion.

CHAPTER 13

PLANNING OF MISSIONS

Although the underground avoids fighting when success is uncertain, some calculated risks must be taken to insure the continued growth of the organization. The underground, therefore, relies on careful planning to maximize success and minimize risk. As a Vietminh training manual states:

> One must study the situation in the opposing camp, that is to say one must try to find out the dispositions of the enemy, the attitude of the cadres and the combatants towards one another, their morale and their fighting value. Follow closely the opponent's activities: the deployment of his intelligence service, transport, relief of pickets, all give evidence of his intentions and so make it possible to plan operations with every chance of success.[1]

The manual goes on to point out the importance of determining the enemy's weak point—morale, provisioning, laxity—and states that "if the enemy protects himself carefully without presenting any weak points, we must create them before attacking."

PREPLANNING

CONTINGENCY PLANS

In planning attacks, the underground develops contingency plans which can be implemented in the event the original plan proves impracticable. Thus, the failure of one action taken in the course of a mission would not necessarily jeopardize the security of individuals or compromise the mission itself or the underground organization.

The need for contingency planning is illustrated by the 1943 Otto Skorzeny attempt to kidnap Mussolini from an Italian mountain resort hotel. This stronghold on a jagged plateau in the Gran Sasso mountain region was 6,000 feet above sea level and considered invulnerable. There would be no chance for a second mission should the first fail. Therefore, on Hitler's instructions and based on limited intelligence, three plans were devised. Plan A called for a lightning air attack on the nearby Italian airbase in the valley. Minutes after the attack, three German transport planes were to land there, one of which would pick up Mussolini while the other two diverted groundfire. If this plan failed, Plan B provided for a small plane to land in the meadows in the valley directly below the stronghold, pick up Mussolini, and transport him to Rome. In Plan C, the most dangerous of the three, a light plane would land directly on the hotel grounds.

In the actual operation, Skorzeny could not make radio contact with his headquarters in Rome to give the go-ahead for Plan A. Plan B automatically went into effect, but failed because the plane was damaged while landing. Plan C finally succeeded, and Mussolini was delivered to the German headquarters in Rome.[2]

REHEARSAL

A second method for increasing the probability of success is to engage in extensive rehearsal for the mission. The success of this method depends to a large extent on adequate intelligence. In preparing a raid, intelligence information enables the underground to make a mockup of the military installation to be attacked.

All moves are preplanned. In the Philippines, on the basis of intelligence reports, a shopping list of certain items to be collected from the installation during the attack was prepared and issued to the members of the raiding party.[3] In this way they were assured that the required items would be taken in the shortest possible time.

Enemy countermeasures and actions needed to cope with them must also be considered, planned, and rehearsed. Small attacks may be launched against a future target to determine where the reaction force will come from and how long it will take to arrive. This information is used to plan the timing requirements of the mission, the placement of blocking units to ambush reinforcements, and escape-and-evasion routes.

EXPLOITATION OF VULNERABILITIES

The vulnerabilities of the enemy can be determined by observing the established pattern of his trained troops. The underground knows that highly trained troops respond in a given manner and attempts to capitalize upon these patterns. The underground also looks for overconfidence on the part of regular troops, knowing that the most vulnerable part of an enemy system may be that which appears to be invulnerable.

Communications systems can also produce vulnerabilities. There is a tendency to centralize most communications systems so that lateral communications between different military organizations are not as rapid as vertical communications within a single organization. Since requests for help usually go to the higher commands rather than to units of another command which may be physically closer, inordinate time delays usually result.* It is these interface problems which the underground planner seeks to exploit.[4]

On the other hand, if government units have an especially efficient communications and response system, insurgents may capitalize upon this efficiency: they can divert government units by staging diversionary attacks or exhaust them through numerous false alarms or small raids on distant outposts.

Several distinct techniques have been used by insurgents to exploit vulnerabilities: infiltration, surprise, deception and diversion, and creation of fatigue through continuous harassment and provocation.

---

* In one Vietnamese province Viet Cong attacks were successful because the zone was situated between two corps of government troops who were unsure of their jurisdiction, and on the border of two provinces whose provincial chiefs suffered the same jurisdictional confusion. Poorly defined jurisdictional and communication responsibilities severely handicapped rapid response by government forces.[5]

Infiltration

Prior to a scheduled raid, underground agents infiltrate an area and locate sympathizers who provide information about the target and government security forces. Frequently, infiltrators devise a screen of lookouts and a system of signals. The underground prepares the battlefield by moving in supplies and arms for use by units which will infiltrate later.[6] If the target is a strategic hamlet, previously infiltrated members of the underground disrupt internal defenses at the time of the raid. They may cut holes in the barricade or barbed wire, move machineguns to prevent overlapping fields of fire, or sabotage weapons arsenals.

Surprise

According to one unconventional warfare expert, it is rarely possible to actually surprise a well-trained enemy through a single motion, such as catching him asleep. Instead, the key to successful surprise attacks is to retain the initiative. This can be done through a three-phase surprise operation: 1) Initiate action against the enemy; 2) allow the enemy to fully develop his reaction; and 3) when the enemy's entire attention is devoted to the reaction, take advantage of his deployment and deliver a surprise counterattack. For example, if the enemy were in an installation with prepared defenses, its reaction must be foreseen. One tactic is to send a provocative force to within three miles of the enemy and set up a dummy bivouac. To draw attention to themselves, this force sends a small patrol which, after encountering the enemy, retreats to the bivouac. As enemy troops rapidly pursue them, believing them to be the only force, the remaining elements prepare an ambush. The principle here is that, although it is not normally possible to surprise and defeat well-trained and well-prepared troops, by diverting their attention to a false target they can be made vulnerable to attack or ambush.[7] In South Vietnam, for example, one of the favorite plans of the Viet Cong is to have a small force attack a village while a large force waits to ambush the mobile reaction troops which rush to the rescue of the village.

Uncertainty is often a key element of surprise attack. Skorzeny found that even well-trained troops are vulnerable for a short time to direct surprise attack if the attack force can create a condition of uncertainty and confusion among the defenders. In a raid of this type, the attack force needs only a small amount of time to take advantage of the confusion.

In the Gran Sasso raid, for example, Skorzeny's troops were briefed to emphasize speed and timing. Skorzeny calculated that only three minutes of hesitation on the part of the defenders would allow his forces to gain the advantage and secure the target. To achieve this hesitation and confusion, a man in the uniform of an Italian general was used as a member of the raiding party. His presence and his shouting of orders distracted and confused the Italian guards long enough so that the garrison was taken without a shot being fired.[8]

During World War II, a Danish sabotage team informed a factory's security force of an expected sabotage raid on the installation, then saboteurs in regular police uniforms entered the factory on the pretext of being reinforcement guards, quickly assembled the entire guard force for an emergency briefing and disarmed and detained them.[9]

Well-trained soldiers are not likely to be surprised or caught off guard when faced with conventional situations, but in situations so uncertain or so unusual that they do not have a prepared response, they can be confused into inaction. This inaction can be exploited, or soldiers can be made to follow the orders and direction of someone who appears to have authority.

Deception and Diversion

To cover a major attack, various means of deception and diversion are used. The insurgents sometimes distract reinforcements or protective forces by attacking other locations or assembling large crowds. In Malaya small attacks and withdrawals were used to make the enemy believe that an attacking force had been driven away. Then a sudden, large attack caught the defenders off guard.[10]

Harassment and Provocation

Continuous harassment by means of hit-and-run attacks on distant villages can fatigue and reduce the morale of well-trained mobile reaction troops. The Vietminh manual states:

> We resort to diversionary and harassing attacks in order to disturb and wear out the enemy; we encircle and split up his positions; we disrupt his lines of communication in order to make him come out to repair them, and then we attack him. In the course of battle we give the impression that we are forced to withdraw, so that the enemy takes up the pursuit and reaches our positions, where we wipe him out. We use stratagems and provocations to lead him on.[11]

OTHER CONSIDERATIONS

Human Factors

In planning raids or other attacks on enemy installations, certain human factors are given special consideration. An attack has greater chance for success if the government morale is low, or if government forces are overconfident or alone, with no available assistance.[12]

Situational factors, such as time of day, are also significant. Night, for example, provides natural cover to prepare for a dawn attack. Adverse weather, such as heavy rain or fog, provides an excellent opportunity for attack. Holidays, weekends, fiestas, and paydays catch the protecting force at a psychological disadvantage.[13]

Tactical Considerations

All things being equal, the smaller the team the greater the security. Usually small teams are selected to carry out ambushes or raids. It is impossible to anticipate all circumstances, so each man must be fully briefed on the mission's goals and prepared to take initiative when appropriate.

The tactical principle of temporary superiority of force predicates the entire concept of small-unit actions. Mao Tse-tung said that strategy was one against ten and tactics ten against one—that a small body can gain temporary advantage over any larger unit. What the underground lacks in strength it seeks to compensate for by the utilization of surprise and rapid action.

Withdrawal Plans

It is most important to plan for a safe withdrawal. A standard part of attack plans is the use of a blocking unit to hamper enemy reinforcements. During withdrawal, these blocking units are used to ambush pursuing enemy forces and enable the main group to escape.[14]

During the withdrawal, casualties are carried away to prevent their identification and reduce intelligence leads for the security forces. Withdrawal is made over a route different from that used to enter in order to avoid ambushes.

For example, in Malaya, blocking units were set up to engage enemy reinforcements, to destroy communications such as bridges and radios, and to counterattack; they left the village last and engaged any counterforces. Their exfiltration routes varied and they rendezvoused after withdrawing from the area.[15]

Review of Mission

After completion, unit leaders review the mission as to successes and failures, sometimes forwarding a detailed report to the higher echelons of the underground organization.[16]

INTELLIGENCE

To plan and organize its activities an underground requires intelligence information. Effective evaluation of intelligence makes it possible to establish priorities among enemy targets and to expose, create, and take advantage of vulnerabilities. In planning psychological operations, intelligence can reveal the attitudes, grievances, and specific problems of a target group so that propaganda themes and agitation slogans can be appropriately developed. Intelligence information regarding trails, tunnels, safe areas and location and capability of enemy forces is essential for evasion operations and escape networks.

ORGANIZATION

One of the first tasks of an underground movement is to establish an intelligence network. The basic unit is the three-to-seven-man cell. For security reasons, cell members never learn the real names or addresses of other members and never come in contact with them.[17] An agent gathers intelligence and transmits it to the cell leader through a courier or maildrop. He never contacts other agents and contacts the cell leader only through intermediaries. Lateral communications and coordination with other cells or with guerrilla units operate on this fail-safe principle so that the compromise of one unit does not jeopardize the security of other units.[18]

Elaborate intelligence systems have been developed to carry on the underground intelligence functions. In Vietnam, for example, the Hanoi-based Central Research Agency directs Viet Cong intelligence. In its headquarters there are six sections responsible for administration, cadres, communications, espionage, research, and training. These sections are responsible for specialized activities within their purview. For instance, the research section has subsections which deal with political, economic, and military affairs.[19]

This agency coordinates the recruiting, training, and dispatching of underground intelligence agents and teams in South Vietnam. A 19-year-old Vietnamese youth, captured by government forces on his first mission, told a characteristic tale. He had been recruited, given a special political training course, and dispatched to infiltrate student groups in South Vietnam. To do this, he was to pose as a defector, move in with relatives living in South Vietnam, and enroll in school. Specifically he was to observe his fellow students, study their personalities, capabilities and aspirations, collect biographical data on them, befriend potential recruits, and report regularly to his cell leader.[20]

The organization and functions of the North Korean intelligence net, set up by zones in the northern half of the Republic of Korea, offer an elaborate and detailed example of how an intelligence net operates in a hostile environment. Its mission was to feel the pulse of the people. The net consisted of 3-member units. Three or more of these units covered a district, and ten or more units a province. Before being accepted, every individual underwent a screening process. Each agent had appropriate identification papers, code devices, and maps indicating places in which to make contacts. While operating in the South Korean Republic, he was to pretend to be a supporter of the government and to mingle with government sympathizers in the villages. Once accepted as such, he was to infiltrate political parties and military units to organize intelligence operations.

The information targets assigned to these intelligence cells included: 1) public opinion about the South Korean Government; 2) attitudes of political parties, social organizations, and governmental agencies toward the public; 3) intelligence activities of the South Korean Government, procedures of dispatching espionage agents, and names of people who were important in political and intelligence agencies; 4) intelligence on military units; 5) strategic intelligence on imports and exports; and finally, 6) party strife.

Procedures for the delivery of intelligence to districts and provinces were also prescribed. If necessary, messages were hidden on the person and destroyed if there was danger of capture. Code words were used to identify messengers, and security measures were devised to protect messengers and the intelligence they carried. The political security chief was responsible for protecting the families of agents so that captured agents would not under coercion compromise the net.[21]

In Algeria, the FLN organized a rudimentary but effective intelligence net. It posted civilian auxiliaries to act as agents in the field. These auxiliaries infiltrated French-held villages, reconnoitered for guerrilla columns. They provided to the liaison intelligence officers of nearby units a steady flow of intelligence about such things as number of French troops, types of armament, and probable targets.[22]

SOURCES OF INTELLIGENCE INFORMATION

In typical underground operations an intelligence screen of two concentric perimeters is established around an area of operations. The outer perimeter usually consists of "innocents"—old men, women, and children—while the inner perimeter comprises members of the unit. When those in the outer perimeter spot an advancing government patrol or helicopter, they alert the inner perimeter by some predetermined signal. Those in the inner perimeter in turn alert the unit.[23]

The Huks in the Philippines depended upon the villagers for intelligence information and improvised techniques to relay this information. If government troops approached a village and a man chopping wood observed them, he would increase the rate of his swing. A woman noticing his increased pace would place white and blue dresses side-by-side on the clothesline. Other members of the security net would pass the warning on that a government patrol was in the area. At night, the Huks used light signals, such as opened windows on a certain side of the house. Each improvised signal always blended with the environment.[24]

In Malaya, rubber tappers relayed intelligence to the Communist terrorists through signals. Pieces of strings hanging from trees or sticks leaning against trees indicated the presence or absence of security forces.[25]

In South Vietnam, the Communist Viet Cong set up an observation system to warn of the takeoff and impending raids of helicopters. In some cases underground echo chambers have been constructed to listen for approaching aircraft. Other agents watched for aircraft arriving in a target area and sounded the alarm and opened fire as the helicopters descended.[26]

Women play an important intelligence role because they appear to be "innocents" and because they can often get jobs in military installations as secretaries or in the homes of enemy personnel as servants.[27]

People in certain professions have direct access to valuable information while carrying out their jobs and are particularly useful to the underground. Professions or jobs which require travel such as shipping, railroads, or trucking provide excellent covers for underground work. The French underground used a doctor as an agent. He made house calls within the German defense area, collecting pertinent information and passing it on to the net through "prescriptions" delivered to pharmacists. To gather intelligence on the coastline, the French underground organized a peat-collecting company which, having access to geologic maps, was able to plot German defense installations and observe the type of construction.[28]

A German innkeeper and a British agent conducted another successful net. An inn located on the Kiel Canal near the Baltic Sea was the favorite hangout of German submariners. The innkeeper made every visit of the German submariners a big occasion and talked the guests into signing the guest register before their departure to sea. He then delivered the register to a British agent who sent the names of the submariners to the British Naval Intelligence. In this way, the British knew the name and departure time of each embarking submarine.[29]

One valuable, though often overlooked, source of intelligence lies in open journals and newspapers. In 1935, the German journalist Berthold Jacob shocked the German intelligence agencies by publishing a book about the German Army, which was then in the initial stage of Nazi rearmament in violation of the Versailles Treaty. In his book, Jacob spelled out "virtually every detail of the organization of Hitler's new army, . . . the command structure, the personnel of the revived General Staff, the army group commanders, the various military districts, . . . the names of the 168 commanding generals and their biographical sketches."[30] Jacob had pieced together scraps of information from obituary notices, wedding announcements, criminal reports, and other such items and eventually compiled a comprehensive picture of the growing German military establishment. It was, as the German authorities agreed, a masterful job of professional intelligence. Hitler's aide reported at the end of the investigation: "This Jacob had no accomplice, My Fuehrer, except our own military journals and the daily press."[31]

In another instance, an article published in a Japanese technical journal revealed details of the new and hitherto secret American U-2 reconnaissance airplane. The article raised questions in the United States as to the possibility of a breach in security. Investigation revealed, however, that the article was simply the result of a painstaking assembly of small items published in various U.S. journals.[32]

In contrast to the highly decentralized underground intelligence organization, the guerrilla force may develop a highly sophisticated, centralized intelligence organization. For example, the Vietminh in Indochina organized the Quan Bao (Military Intelligence) in 1948, to provide operational intelligence for its forces. The Quan Bao was responsible for the collection and coordination of all military intelligence. The Quan Bao was established as an elite corps and was made up of party members selected on the basis of their physical, mental, and moral qualifications.

The personnel selected went through a 12-week training course which emphasized physical conditioning, self-defense, techniques of sensory perception and memory, background information on the French, reconnaissance and interrogation techniques, and methods of accurate and complete reporting and evaluating of incidents and situations.

Special emphasis was placed on methods for obtaining prisoners. Units made up for this specific purpose had four subsections: (a) a fire group to create confusion in the enemy ranks; (b) a capture group to round up prisoners; (c) a support group to assist in retention of prisoners and to watch for enemy reinforcements; and (d) an escort group to take prisoners to an interrogation area.

Various methods were employed to elicit information from the prisoners. Though physical torture was seldom used, Vietminh agents interrogated individuals for long periods at times when the prisoner's resistance tended to be lowest. Sarcasm and irony were often used to make the prisoner lose patience and composure. Vietminh agents sometimes disguised themselves as prisoners and mingled with other prisoners in the compound.

At company and battalion level *trinh sat* sections were responsible for reconnaissance and security. In areas which were considered suspect, agents were assigned to keep the units supplied with information. They investigated areas of operation, reconnoitered possible ambush positions, and determined ingress and egress routes. After a combat mission, the agents led the troops back to a regrouping area, took over prisoners, and authorized the local civilians to return to their homes. A report of the operation, complete with tallies of casualties suffered and inflicted, weapons lost and captured, and an evaluation of the unit's performance and mistakes in action, was made.

In the camp areas and villages, agents maintained close surveillance over Vietminh troops to guard against desertions and to insure that troop behavior would not damage relations with the local population. The agents were responsible for the security of documents and automatic weapons. They also maintained perimeter lookout positions to warn of enemy aircraft, and directed emergency evacuation procedures.

The efficiency and scope of the Quan Bao were revealed through captured documents which contained highly detailed and accurate surveys of French troop dispositions, habits, and activities. Surveys of areas of French operations included terrain trafficability for both vehicles and coolies, as well as loyalty and attitude estimates of nearby native populations.[33]

FOOTNOTES

[1] Otto Heilbrunn, *Partisan Warfare* (New York: Praeger, 1962), pp. 80–81.

[2] Otto Skorzeny, *Secret Missions* (New York: E. P. Dutton, 1950), pp. 100–103.

[3] William J. Pomeroy, *The Forest: A Personal Record of the Huk Guerrilla Struggle in the Philippines* (New York: International Publishers, 1963), p. 26.

[4] Edwin F. Black, "The Problems of Counter-Insurgency," *United States Naval Institute Proceedings*, LXXXVII, No. 10 (October 1962), pp. 22–39.

[5] "Viet Cong Control Main Roads, Use Tolls to Finance Raids," *The Evening Star* (Washington, D.C.), (February 11, 1965), p. A–3.

[6] "Viet Rebel Tells of Tactics Used Earlier at Bien Hoa," *The Washington Post*, February 8, 1965, p. A–10; Fred H. Barton, *Salient Operational Aspects of Paramilitary Warfare in Three Asian Areas*, ORO–T–228 (Chevy Chase, Md.: Operations Research Office, 1953), pp. 200 and 217; and George K. Tanham, *Communist Revolutionary Warfare* (New York: Praeger, 1961), p. 25.

[7] Otto Heilbrunn, *Warfare in the Enemy's Rear* (New York: Praeger, 1963), pp. 215–16.

[8] Burton F. Hood, "The Gran Sasso Raid," *Military Review*, XXXVIII (February 1959), pp. 55–61.

[9] Lt. Jens Lillelund, "The Sabotage in Denmark," *Denmark During the German Occupation*, ed. Borge Outze (Copenhagen: The Scandinavian Publishing Co., 1946), pp. 53–56.

[10] Barton, *Paramilitary Warfare*, p. 200.

[11] Heilbrunn, *Partisan Warfare*, p. 81.

[12] Barton, *Paramilitary Warfare*, p. 204.

[13] *Ibid.*, pp. 200 and 204.

[14] *Ibid.*, p. 200; also Heilbrunn, *Partisan Warfare*, p. 94.

[15] Barton, *Paramilitary Warfare*, pp. 200–201.

[16] Lillelund, "Sabotage in Denmark," pp. 53–54; see also U.S., Department of State, *A Threat to the Peace: North Vietnam's Effort to Conquer South Vietnam*, Part I (Washington, D.C.: Department of State, 1961), p. 102.

[17] Intracell behavior is described in the following sources: U.S., War Department, Special Staff, Historical Division (Historical Manuscript File), *French Forces of the Interior, 1944* (Washington, D.C.: General Services Administration, Federal Records Center, Military Records Branch), p. 320; Ronald Seth, *The Undaunted: The Story of Resistance in Western Europe* (New York: Philosophical Library, 1956), p. 259; and Jon B. Jansen and Stefan Weyl, *The Silent War* (New York: Lippincott, 1943), p. 115.

Communications among members of an intelligence cell are discussed in Alexander Foote, *Handbook for Spies* (London: Museum Press, 1949), p. 44; also Fred H. Barton, *North Korean Propaganda to South Koreans (Civilians and Military)*, Technical Memorandum ORO-T-10 (EUSAK) (Chevy Chase, Md.: Operations Research Office, 1951), pp. 151–57.

The regulation of vertical liaison is discussed in Gershon Rivlin, "Some Aspects of Clandestine Arms Production and Arms Smuggling," *Inspection for Disarmament*, ed. Seymour Melman (New York: Columbia University Press, 1958), pp. 191–202; Jan Karski, *The Story of a Soviet State* (Boston: Houghton Mifflin, 1944), p. 160.

[18] Andrew R. Molnar, *et al.*, *Undergrounds in Insurgent, Revolutionary and Resistance Warfare* (Washington, D.C.: Special Operations Research Office, 1963), pp. 53–54.

[19] U.S., Department of State, *Aggression from the North: The Record of North Vietnam's Campaign to Conquer South Vietnam* (Washington, D.C.: Department of State, 1965), p. 25.

[20] *Ibid.*, pp. 12–14.

[21] Barton, *North Korean Propaganda*, pp. 151–55.

[22] Paul A. Jureidini, *Case Studies in Insurgency and Revolutionary Warfare: Algeria 1954–1962* (Washington, D.C.: Special Operations Research Office, 1963), p. 100.

[23] Luis Taruc, *Born of the People* (New York: International Publishers, 1953), p. 121; and "Fighting the Viet Cong," *The Evening Star* (Washington, D.C.), May 12, 1965, p. A-16.

[24] Col. Napoleon Valeriano, speech at the Counterinsurgency Officers Course, Special Warfare School, Ft. Bragg, N.C., November 5, 1963.

[25] Richard L. Clutterbuck, "The Cold War," *The Army Quarterly and Defence Journal*, LXXX, No. 2 (January 1961), p. 166.

[26] "Viet Cong Guerrilla Tactics Are Hit, Run and Hide," *The New York Times*, July 2, 1965, p. 2.

[27] Soviet Army, *Handbook for Partisans* (Moscow: Field Service Regulations, Soviet Army, 1942). Translated in Alexander Pronin, *Vostok* (Washington, D.C.: Operations and Policy Research, Inc., n.d.), p. 67.

[28] For a complete narrative of the story of the French underground's intelligence work in charting the Germans' "Atlantic Wall" defenses, see Richard Collier, *Ten Thousand Eyes* (New York: E. P. Dutton, 1958).

[29] *Ibid.*

[30] Ladislas Farago, *War of Wits* (New York: Funk and Wagnalls, 1954), p. 55.

[31] *Ibid.*, p. 57.

[32] Margret Boveri, *Treason in the Twentieth Century* (New York: G. P. Putnam's Sons, 1963), pp. 330–31.

[33] George K. Tanham, *Communist Revolutionary Warfare* (New York: Praeger, 1961), pp. 70–83.

CHAPTER 14

OPERATIONS

AMBUSHES AND RAIDS

AMBUSHES

Ambush has been defined as a surprise attack upon a moving or temporarily halted enemy with the mission of destroying or capturing his forces. The surprise attack is usually from a concealed position and with sudden concentrated fire. Maximum effect is achieved when the ambushed force does not see the ambushing force and its area of movement is limited. The ambushing force usually has the advantage of a short field of fire and covered routes of withdrawal. Accordingly, a column of troops moving along a narrow jungle road is a prime target. An effective ambush is usually based on advanced intelligence and detailed planning, and executed with imagination and boldness.[1]

Purposes of Ambush

Ambush is a tactic which is used extensively.[a] It plays a part in 60 to 70 percent of Communist combat actions.[3] It is an effective means of acquiring weapons, harassing and demoralizing government forces, delaying or blocking the movement of troops and supplies, destroying or capturing government troops (especially government officials and army officers), and undermining confidence among the populace toward the government forces.

Capturing Weapons and Supplies. Ernesto "Che" Guevara noted that the foremost aim of an ambush is the acquisition of weapons; only in special situations should attacks be made without the prospect of capturing weapons.[4]

The Viet Cong has used ambush frequently for this purpose. In two ambushes in February 1964 they captured a total of 166 weapons. During the previous 8 months, the Viet Cong had inflicted other heavy weapons losses on government troops and had captured American-made radios and were able to monitor the government's communications network.[5]

Blocking Movement of Troops and Supplies. The Vietminh in 1950–54 used the ambush effectively against French relief units. The Vietminh's customary tactic was to split their force into three groups and position them at strategic positions along the road where the relief column was expected to move. The first group blocked the road while the second and third groups hid at separate places along the approach road. After the French

---

[a] In the Philippines the Huks ambushed the armed forces of the Philippines 1,864 times out of 2,145 armed engagements between April 1950 and January 1952. In South Korea, the National Police reported that out of 2,868 contacts during a 3-month period in 1958, 1,886 were classified as ambushes or raids. During the same period in Malaya, authorities recorded 1,890 ambushes out of 3,133 major incidents.[2]

convoy passed the hidden second and third groups they encountered a roadblock. As the French got out of their vehicles to inspect the blockade, the first and second groups attacked while the third sealed off the rear, preventing retreat or the arrival of help.[6]

Diversionary Use of Ambush. The ambush is also a widely used diversionary tactic. For example, underground units in danger of being encircled by security forces may ambush a portion of the encircling troops, diverting the attention of the security forces from the encircling operation. In addition, an ambush can attract security forces' attention long enough for numbers of troops to successfully cross important highways or other strategic areas.[7]

Other Uses. A hit-and-run ambush tactic is generally used for harassment, capture, or destruction of enemy troops. In Vietnam, for example, a minor diversion may be created on a road near a security force camp by burning a passenger bus or stealing identity cards from civilians. Receiving information of the attack, the security forces proceed to the place of attack and are ambushed on their way while still in their vehicles.[8] Occasionally the Viet Cong use an attack on a government post as a decoy. The defenders ask for help while a much larger guerrilla force prepares to ambush the relief column on the anticipated route.[9]

Ambush is also used to capture enemy personnel for interrogation [10] or to kidnap or assassinate top government officials and army officers. In another variation, undergraduates dressed as government soldiers may ambush and rob innocent civilians to turn the populace against the government. The Vietminh guerrillas used ambush in an attempt to paralyze the French communications by denying them the use of roads, paths, and waterways.[11]

Government forces have, of course, found that the ambush can be an effective counterguerrilla tactic. British commanders in Malaya found it to be the most successful tactic used against the Communist terrorists.[12]

Ambush Tactics

The basic tactic of ambush is the use of the smallest possible number of men employing the military principle of surprise and avoiding open combat with numerically superior forces.

An analysis of ambushes in Malaya, Korea, and the Philippines indicates that the ambushers were able to lay down a continued and effective line of fire against superior government forces, while maintaining their capability to disperse quickly. Adeptness at dispersal and withdrawal depends on optimal timing and placement of ambushers. Concealed attacks on main supply routes were the most frequent kinds of ambush used in Korea, Malaya, and the Philippines. Of 82 ambushes recorded in 1951, 62 occurred along main roads, 14 against patrols in hills or jungles, and 6 in small villages. Instances of urban ambush consisted mostly of hit-and-run attacks or holdups of public conveyances, executed by groups of three to five men.[13]

In Malaya, motor transports were ambushed with impunity; vehicles were easy targets since the few available roads were winding, hilly, and cut through thick vegetation and narrow gorges. Most of the ambushes against government forces in Malaya occurred while they were moving through dense jungles where the attackers had the tactical advantage of concealment and close-range firing. In Korea, ambushers used the same tactics although the roads were not foliaged, but had boulder outcroppings for hiding places. Convoys and patrols were frequently ambushed in mountain passes where the road was cut through rock defiles.[14]

In Southeast Asia, ambushers frequently camouflage hardened bamboo stakes along the trails. When the enemy is ambushed by automatic-weapons fire and dive for cover in the undergrowth, they are impaled on the hidden stakes. The bayonet-like stakes can inflict as many casualties as the weapons fire.[15] In South Vietnam, the Viet Cong strew hidden stakes along the trails to prevent rapid pursuit of ambushers by government forces after an attack.

A frequently used formation is the L-ambush (see figure 9), in which part of the ambush party is placed in front of the enemy and part on his flank. As the enemy approaches he is fired upon from both front and flank. This

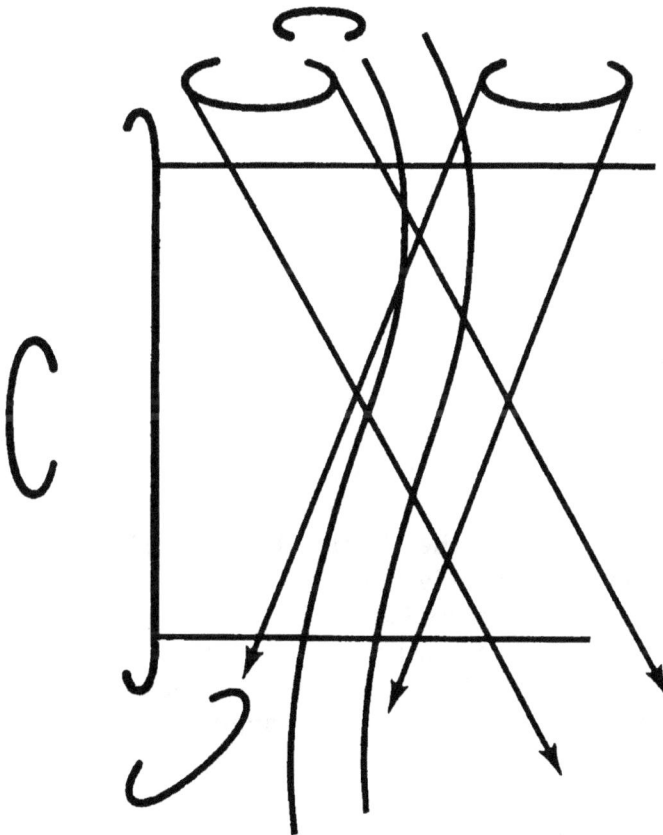

L-SHAPED AMBUSH

Figure 9. L-shaped ambush.

positioning requires only one route of withdrawal and permits the unrestricted use of automatic weapons. If the enemy were encircled in the ambush, the use of automatic weapons would be restricted and more than one route of withdrawal would be required. The V-type is another form of ambush which takes advantage of fields of fire that place the column in a crossfire [16] (see figure 10).

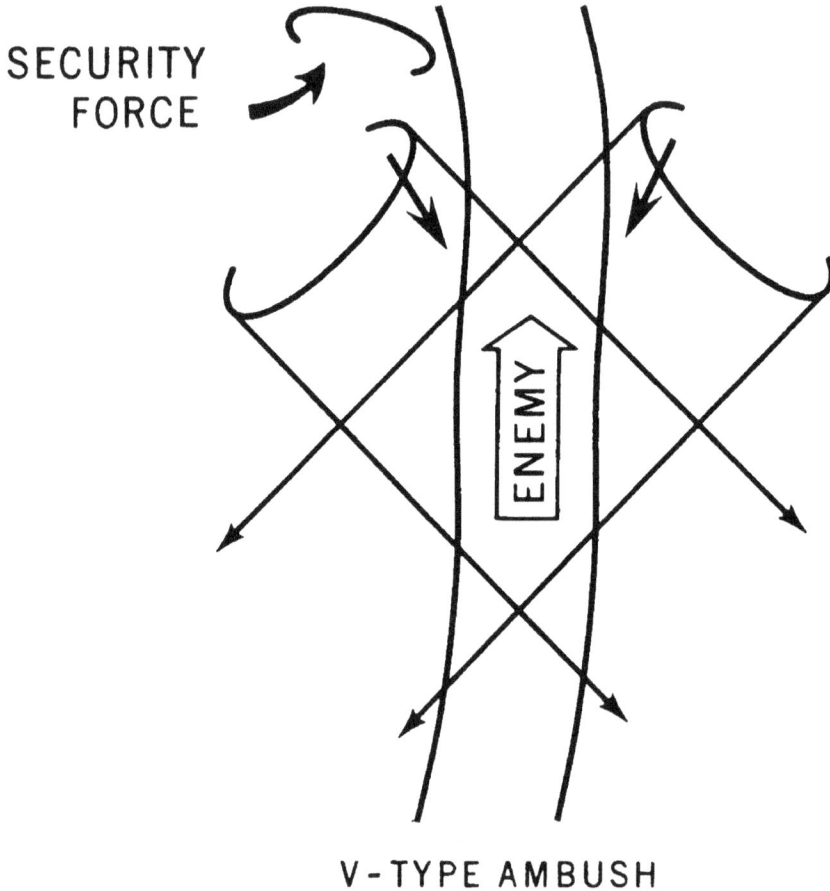

SECURITY
FORCE

ENEMY

V-TYPE AMBUSH

Figure 10. V-type ambush.

Guevara recommends another tactic, the "ambush and vanguard." He notes that an enemy column advancing through an area of thick vegetation or woods is unable to secure its flanks. To guard against possible ambush, the government forces usually send out a reconnaissance vanguard column, which provides point-and-flank security by probing for hostile forces. However, by advertising its presence, the vanguard inevitably exposes itself to danger. Guevara suggests that the ambush be set for this vanguard: when the vanguard reaches an agreed-upon point, preferably at the highest terrain point, then the ambushers open fire. While a small force holds the main column back, arms, ammunition, and equipment are collected from the vanguard unit and then the ambushers quickly withdraw. Guevara believes that an attack on vanguard units is an ideal operation for small forces.[17]

It has frequently been claimed that the ambush is responsible for most of the casualties of government forces. It is estimated that in Korea ambushes inflicted 55 percent of the government casualties; in Malaya, 75 percent; and in the Philippines, 60 percent. Yet the ambushers, since they were on the offensive, suffered comparatively few casualties in the ambush operations. Total casualties among the ambushers, including those killed, wounded, and captured, were only 15 to 20 percent in Korea and the Philippines and 10 percent in Malaya. By inflicting heavy casualties on the enemy at small risk, ambush operations proved to be an extremely effective tactic.[18]

Ambush Countermeasures and Human Factors

Vietminh tactics eluded most French attempts to clear out ambushing forces. Vietminh manuals emphasized that encircling forces were never strong at all points and that a forceful and sustained attack could open an avenue of escape. It directed an ambushing force, caught in a tight encirclement, to concentrate all fire on the weakest point of the surrounding force in order to fight its way out. The manual also advised the ambushing force to withdraw if danger of encirclement was probable.[19]

Reviewing their operations in Vietnam, French officials have plaintively observed that, in spite of air and other techniques of observation, they were consistently unable to detect ambushes in advance. They found that, at best, their defenses were only able to lessen the effectiveness of ambushes, not forestall them. One countermeasure employed was to extend convoys and columns for a longer distance than could possible be covered by an ambushing force; thus part of the force could escape, while the other part could come to the rescue. Another technique was to equip columns with armor and artillery in order to assist in repulsing ambushers.[20]

U.S. Army advisers have suggested five methods of countering ambushes in Vietnam.

1. Government forces must not follow consistent patterns of movement. Any one road should be used infrequently, and the timing and size of the road patrol varied. Also, the security patrol should carefully check out suspected ambush sites.

2. Reaction of troops to an ambush must be fast and automatic.

3. Effective communications must be maintained with the base.

4. The ambushers must be pursued aggressively and not simply permitted to withdraw.

5. Village activity must be observed. Security forces should interpret such signs as an empty marketplace as notice of impending action.[21]

Another instance of Viet Cong ambush emphasized this final point. The security force did not search a nearby village that was considered loyal. When it passed through, it did not interpret the absence of villagers as a sign of danger; subsequently, the Vietnamese troops were ambushed, the villagers having been bound and gagged by the Viet Cong.[22]

Australian Army research into the human factors involved in ambush countermeasures suggests that only 15 to 25 percent of soldiers faced with sudden danger respond immediately with fixed purpose or effective activity. A majority are stunned and bewildered. The Australian Army, therefore, has developed a special counterambush drill.

The counterambush drill is an attempt to train soldiers to force an ambush—that is, rush through the ambushers and envelop them from the rear. At first it was difficult to persuade soldiers to advance against a concealed position defended with automatic weapons. Even when instructors who had used this tactic in battle reassured them, they did not gain confidence, so a new training technique was devised. The first step was to diagram for the trainee an ambush situation and to stress that the natural reaction is to run away after the first contact. Training, however, conditions the soldier to be positive and aggressive and not to flee from or attempt to bypass an ambush. They then learned that speed was essential to their self-preservation, and that the quickest way to extricate themselves from the ambush was to concentrate their fire and run through a single point along the ambusher's line. In time a marked change in the attitudes occurred. Trainees no longer faced ambushes with fear and adopted instead a tough-aggressive posture. They also developed a pride in their fearlessness.[23]

RAIDS

The raid is a sudden attack against a stationary enemy force or installation. It is characterized by secret movement, brief but concentrated combat action, rapid disengagement, and swift withdrawal.

The purposes of the raid are: (1) to destroy or damage supplies, equipment, or installations; (2) to capture supplies, equipment, or key personnel; (3) to inflict casualties on the enemy and his supporters; and (4) to harass or demoralize the enemy.

As in ambush, a key to success in raids is surprise: to attack the enemy when and where he is least prepared, and to take advantage of weather, visibility, terrain, and other environmental factors.

A study of raids in Malaya, Korea, and the Philippines indicates that raid techniques followed a definite pattern: the majority were aimed at civilian villages and military or police outposts. Village raids in Korea generally sought to terrorize the people. In the Philippines, the Huks conducted raids primarily to destroy military posts and such guarded targets as railroad trains and powerplants, as well as to terrorize village inhabitants, kill enemy forces, and destroy support facilities.[24]

Raiding units were usually divided into two groups: one to secure the approach and withdrawal routes, and one to accomplish the raid mission. In raiding villages, the raiders usually withdrew along a different route from the one they approached on, thereby facilitating escape and permitting the group that secured the approach to be mobile. By moving through hills and jungles, the raiders outmaneuvered police posts on roads leading to villages. The raiders generally were able to infiltrate guarded villages and to disarm the police by surprise. Available casualty reports indicate that in seven out of ten raids the raiders had fewer casualties than the government forces.[25]

In South Vietnam, the Viet Cong have frequently used raid tactics.[b] In January 1964, for example, they raided the village of Thuah Dao, 15 miles southwest of Saigon, in a typical fashion. The guerrillas killed the head of the militia, set fire to the village administrative office, and forced the villagers to tear their homes apart. Forty homes were destroyed while the terrified militia radioed nearby units for help to no avail; demoralized, the militiamen surrendered their weapons. Such Viet Cong raids have spread terror among local officials, undermined the morale of government soldiers, and have been useful for capturing equipment and supplies. In a raid on a U.S. Special Forces training camp high in Vietnam's central plateau on July 5, 1965, the Viet Cong killed 41 soldiers, seized the camp's arms and ammunition (including a 60-mm mortar, 4 heavy machineguns and more than 100 other weapons), and withdrew before the arrival of government troops.[28]

A typical Viet Cong raid designed to destroy important South Vietnamese equipment occurred shortly after midnight on November 1, 1964.[29] Using American-made ammunition and six 81-mm mortars, the Viet Cong attacked the Bien Hoa air base for 20 minutes. Targets were the control tower, the quarters of military personnel, and a cluster of B–57 aircraft on a parking ramp. Six men were killed and 21 injured, with five B–57 aircraft destroyed and 15 damaged. Despite a counterattack by artillery and aircraft, the Viet Cong were able to escape.

The effectiveness of this attack rested on careful planning and intelligence. Vietnamese who had previously lived in the area had been forced off when the base was constructed. A number of these local villagers reportedly aided the Viet Cong in infiltrating the base area and obtaining enough information to make a life-size model of it. This preparation permitted the mortar squads and covering infantry units to conduct practice raids. Ranges, azimuths, and positions were determined for the mortars to insure direct hits on three prime targets in the minimal amount of time.

The mortars were set in place piece-by-piece by small squads of infiltrators. The actual shelling lasted only 20 minutes because no targeting rounds were necessary. With the infantry blocking unit covering against possible government reaction forces, the mortar squads exfiltrated in small groups with their weapons.

The success of the Bien Hoa raid illustrates a number of human factors involving timing, security, and local attitudes. In executing the attack the Viet Cong worked all of these factors to their advantage. The raid occurred the day after payday, early Sunday morning, following a national holiday celebrating the first anniversary of Diem's overthrow; and it occurred during the monsoon season when combat operations generally cease. Further, the raid coincided with the change in guard, during a period

---

[b] One interesting technique the Viet Cong followed during raid operations was to demand that villagers kill all their dogs so that their barking would not betray the guerrillas when they made subsequent raids.[26]

of uncertainty, following the installation of a new government in nearby Saigon. Also the division of responsibility for internal and perimeter security between United States and Vietnamese forces, respectively, was unclear for certain functions such as establishing perimeter patrols and providing reaction troops. The Viet Cong infiltrated the Vietnamese perimeter with mortars and fired into the American zone. Since the attack was in the Vietnamese area, the Americans could not respond, and since the Vietnamese were not being attacked, they did not respond.

The Huk insurgents in the Philippines also capitalized upon certain human factors in planning their raids. Generally raids occurred at night or dawn; daytime raids took place only in isolated areas or where some feature of weather or terrain provided cover or concealment.[30] Raids frequently took place immediately before a holiday festival when most villagers were preoccupied and garrisons weakly defended.

SABOTAGE

Sabotage is an attempt to damage the resources of the government's war effort—military and economic organizations, industrial, food, and commodities production, and public morale and law and order.

In most instances, material damage inflicted on the enemy by sabotage is relatively small. However, in some cases, sabotage by a trained underground team can be more effective and less costly in manpower and material resources than large unit operations or aerial bombings. For instance, the destruction of the German heavy-water plant at Telemark, Norway, was assigned to undergrounders after aerial bombing raids and a glider-borne commando attack had failed.[31]

Organized sabotage attacks do indicate to the population that the underground movement has the will and the strength to perform these acts in spite of the government and its security forces. General sabotage and planned attacks can create temporary disruption of transportation and communications, causing the government or occupying force to garrison more troops for security duty. Finally, sabotage can lower morale among the security forces. Repeated acts can induce fear that such acts will be repeated on a larger scale.[32]

STRATEGIC SABOTAGE

Strategic sabotage involves direct action by specially trained underground units against such key targets as factories and military installations. The units depend on intelligence reports to establish priorities among these targets.

The underground in occupied Denmark established an "industrial council" to compile and analyze information about prospective target installations, and assign priorities for sabotaging efforts.

Underground agents infiltrated targets to contact sympathetic employees and acquired blueprints, diagrams, charts, and defense plans. They then determined the vulnerable points of access, critical and irreparable machinery, guard procedures, searchlights, dogs, and so on. A close surveillance of the installation verified and augmented this data.

The units then planned their attack, taking into account the following factors:

1. The method of entry, which should have an element of surprise, and so requires a thorough understanding of the guard force and security procedures of the target.

2. Amount and type of explosive and its placement considering the access routes, the time involved in placing the explosive and withdrawing from the target, and the timing of the fuse on the explosive.

3. A covering unit, situated to observe ingress and egress routes, to intercept or ambush a pursuing ground force and insure the withdrawal of the sabotaging unit.

After an attack, the group leader made a detailed report of the operation, which he forwarded to officials in the underground organization and to the Allied supporting units in England.[33]

One noted strategic sabotage operation was carried out against a heavy waterplant near the town of Vemork, in Telemark Province, Norway, on February 28, 1943. The Allied forces of Great Britain, Canada, and the United States had acquired information that this plant was valuable to German atomic energy experiments and, to prevent further German progress in this field, the Allies gave high priority to its destruction. First, a Norwegian parachuted into Telemark Province to collect intelligence on vital elements of plant operation such as the number, hours of change, and behaviorial pattern of the German guards; the layout of the plant; and routes of ingress and egress. The undergrounders made contact with the plant's chief engineer, a Norwegian, who gave them information about the floorplans and the location of critical machinery. In order to plan and rehearse their attack, they built an exact model of the heavy waterplant in England.

After nearly a year's planning, an 11-man team infiltrated the area for the actual attack. The team divided into two groups, blocking and demolition. The time for the strike was set for 12:30 a.m. to insure that the off-duty guards would be asleep and to allow the undergrounders 5 hours of darkness to escape.

In the actual attack, the blocking unit forced the entrance and covered the German guard barracks, while the demolition unit entered the plant itself through a cable tunnel and set the charge at a predetermined location. Although the explosion was so small it did not arouse the German garrison, it destroyed key apparatus in the plant and 3,000 pounds of heavy water, nearly half a year's production.[34]

An underground may spare potential targets. During World War II the Yugoslav underground spared three particular ones—vital public utilities, lumber mills, and certain railroads. Utilities were excluded because the underground was dependent upon the support of the population and disruption of such a facility as a water supply would impair, not enhance, the underground's support. Although the lumber produced in the local mills was being used by the Germans, the destruction of these mills would impair the livelihood of the population, not only alienating people but, since much of the underground's funds came from the local population, cutting off the supply of money. Finally, they did not raid those German-operated railroads which were as vital to the underground as to the German forces for communications and transportation.[35] It was probably as effective to mis-route the railroad cars through a change in their manifest as it was to destroy them through sabotage.

GENERAL SABOTAGE

General or nuisance sabotage is closely related to passive resistance in that it requires neither trained sabotage teams nor carefully selected targets. Sabotage acts in this category usually express individual resistance and take the form of noncooperation, such as deliberate slowdowns on factory production lines, or harassment, such as telephoned bomb threats that force the evacuation and search of buildings and plants.[36] In German-occupied Russia during the later stages of World War II, nuisance sabotage and partisan actions against the German soldiers forced them to travel only in groups and to remain on main arteries.[37]

Noncooperation sabotage was used extensively in Belgium during the German occupation. Workers slowed their pace of production, went on strike, and refused to help Germans apprehend rebel patriots; postal workers intercepted letters addressed to the Gestapo. The Belgian underground used camouflaged pamphlets to prompt the patriots to sabotage the German occupation. One such pamphlet was a small booklet containing instructions on how to slow down production in factories and sabotage the mines.[38] Also in Belgium, sharp metal objects called crabs were put in the streets to puncture auto tires. This technique, which virtually halted traffic, was very effective because the only people who had automobiles were German officials and Belgian collaborators.[39] The Polish underground pressured doctors into signing medical certificates stating that certain people were unable to work.[40]

Other techniques included failing to lubricate machines in accordance with maintenance schedules, hiding repair parts, and dropping tools and other foreign objects into moving parts. It has been stated that the 250,000 workers in Italy's metallurgical industries at one time cut production by as much as 16 percent through indirect sabotage.[41]

ESCAPE-AND-EVASION NETS

OBJECTIVES

Escape-and-evasion nets in underground movements are established for the infiltration and exfiltration of an area of operations by underground members or agents. Such a system is needed for clandestine operations in which couriers with messages and funds, organizers, or training instructors must move through government-controlled security areas.

To provide maximum security for leaders of the underground organization, escape and evasion networks incorporate the fail-safe principle. In the planning of raids, sabotage, and intelligence missions, methods of escape-and-evasion or withdrawal are of primary concern. Accordingly, operations planners usually consider alternative ingress and egress routes and establish contingency plans for withdrawal and rendezvous.

ORGANIZATION

Escape Routes

Escape-and-evasion networks usually consist of escape routes and hideouts or "safe houses." There are three general categories of safe houses: the temporary stopover, the emergency hideout, and the permanent refuge.[42]

Couriers and traveling agents use the temporary stopover to facilitate travel. Escapees and persons in danger also use temporary safe houses along escape routes for food, rest, and directions to the next stopover.

An agent who becomes suddenly ill, is wounded, or is sought by the police can use the emergency hideout. Such safe houses are usually private homes of loyal and reliable persons who are supporters of but not identified with the underground movement. Other facilities have also been used as safe houses; in Algeria physicians loyal to the FLN put evaders in the Algiers Municipal Hospital as patients.[43]

The permanent or long-term safe house can be an isolated farm or cabin, a distant encampment, or a location in a nearby nation sympathetic to the underground movement.

In the 1930's the Soviet Comintern utilized extensive auxiliary offices and bases for their agents abroad. The Seamen's and Port Worker's International, for instance, controlled seamen's clubs in every major port in the Western Hemisphere. These clubs served as reporting and relocation bases for agents operating in or traveling through the country. Personnel at these auxiliary bases arranged contacts, passports, cover addresses, and funds for agents. When an agent lost contact with his organization he simply reported to the nearest auxiliary base for food, shelter, funds, and instructions.[44]

In planning riots and demonstrations, the Viet Cong established safe zones in sections of the city where they would store weapons and assemble agitators. They identified shopkeepers and homeowners willing to provide shelter for the demonstrators. The agitators hid in these safe houses until the police had completed their postdemonstration search.[45]

The safe houses along an escape-and-evasion network are usually within one day's travel of each other. The person maintaining the safe house seldom engages in any other subversive activity which might draw attention to him. The underground supplies him with extra food, clothes, and any identification papers or documents as needed. Each person in the route knows how to reach only the next link, and no one person knows the identity or location of every link. Guides generally escort the escapee from one link to the next. The guides meet at a pre-arranged spot half-way between the two safe houses and neither guide knows the location of the other's safe house.

The Viet Cong infiltration process from North Vietnam to South Vietnam illustrates the safe house and fail-safe concepts. After completing training in the North, Communist infiltrators are trucked to the Laotian border just above the demarcation line where they rest for several days before beginning their move southward. An infiltration group usually numbers 40 to 50 men, but once they reach the border they break up into smaller groups. Each man carries a 3 to 5 day supply of food, a first-aid packet, hammock, mosquito netting, and similar items. No one may carry personal papers, letters, or photographs that might be used by the enemy to identify him. The infiltration routes along the Laos-South Vietnam border include way stations. A chain of local guides leads the units along the secret trails. Each guide knows only his own way station and conveys troops to the next way station just as the network conveys escapees between safe houses. Conversation is discouraged in transit and only the leader of the group may speak with the guide. In this manner the network maintains a relative degree of security and is not necessarily compromised if one guide defects or is captured.[46]

Some Communist agents operating in non-Communist countries periodically travel to Communist-controlled areas for obtaining supplies, training, and planning. In order to evade government travel controls to Communist countries and to conceal their operations, these agents often travel first to a neutral country and contact a Communist embassy. Communist agents from the various Latin American countries covertly traveling to Castro Cuba used to go first to Mexico. The Cuban Embassy there reportedly gave the agent papers that served as a visa but were not entered into the agent's passport. He then took a Cubana Airline flight to Cuba under an alias. However, the Government of Mexico has begun to keep extensive records, including photographs of all people traveling to Cuba by commercial means. The Mexican customs authorities make these records available to other governments. Thus, agents now must travel to Cuba via eastern European Communist countries to prevent the recording of their travel.[47]

During World War II, underground escape-and-evasion nets devised some unusual techniques to pass escapees beyond checkpoints. Police members of the net would handcuff escapees and pass them through the checkpoint as prisoners. Underground members also hid escapees in maternity homes until their passage through the escape route could be secured. In one incident, an escapee was passed through a checkpoint by placing him in an ambulance and having him feign insanity.[48]

Escape Techniques

A number of successful prison escape techniques were developed in World War II. Popular techniques included traversing tunnels under prison walls; climbing over walls or cutting barbed wire fences; and impersonating military personnel, local workers, and prison working parties.

Escapes were frequent but escapees were not so successful at evading recapture. A prisoner's best chance to escape is on his first day of imprisonment or soon thereafter. As his physical condition weakens from prison food and routine, his chances of escape diminish. Surveillance and planning are the first steps in effecting an escape. Precautions must be taken to keep conflicting plans from compromising any escape attempt. In one prison, an officer was assigned to coordinate and register all escape plans.[49] In another prison, permission of the senior officer had to be obtained. The officer relayed the escape idea to an escape committee which approved the idea and coordinated its execution. This committee included a representative from each barracks, as well as expert tailors, forgers, and intelligence informants. It also maintained an escape book in which all details on escapes, whether successful or not, were recorded.[50]

Once an escape plan is devised, a long period of preparation and surveillance is necessary. An example illustrates the painstaking planning of an escape attempt. Detailed studies were made of the prison layout and behavior of the guards. The committee observed that one window was a blindspot for sentries on a nearby catwalk and at a gate in a barbed-wide fence. Observations of the guards' behavior patterns revealed the length of the posts they walked, the length of time spent walking, the exact time they were relieved, and the length of time it took to return to the guardhouse, alone or in squad formation. One prisoner was to be disguised as an officer of the guard and two other prisoners as sentries. At a fixed time, the disguised officer would relieve the two legitimate guards with the disguised guards, after which the prisoners would attempt to escape protected by the disguised guards. Critical to the escape act was the time required by the sentry relieved by the disguised officer to walk back to the guardhouse and notice the real officer, react, and rush back to his post. The time was calculated to be three-and-a-half minutes to, at most, four-and-a-half. In this time 20 prisoners could slide down the rope and escape through an opening in the fence.[51]

A number of factors are involved in successful escape-and-evasion techniques. Planning, for example, must be based upon careful observation of the prison environment, of the habits and modus operandi of guards and staff. Successful escape has frequently been dependent upon astute use of disguise, paying careful attention to detail, using the suggestibility of enemy personnel and the element of surprise to advantage.

A common device used in escape is the employment of disguise. For example, at one camp, prisoners skilled in tailoring secretly manufactured a chimney sweep's uniform, complete with top hat, tails, black trousers, and weighting ball and brush. Dressed in this and with a pass reproduced from that of a legitimate chimney sweep, a prisoner was permitted to pass through the gates unquestioned.[52]

In another instance, a prisoner disguised himself as a foreign naval officer, complete with decorations and colorful insignia. Police guards and army officials are not apt to question an unfamiliar but official-looking uniform, especially when papers and letters of introduction (forged) look imposing. In this disguise, the escapee was able to travel in broad daylight on regularly scheduled trains.[53]

Another escapee capitalized upon his guards' respect for authority. Dressed as an army general, he approached a prison gate just after the change of sentry (timed to make the new sentry believe that he had checked in with the previous sentry). The general berated the sentry because his chauffeured automobile was not there. When the sentry said he would call about it, the general told him to forget about it, he would walk.[54]

A similar technique was used on another escape. Two prisoners wearing the overcoats of officers approached the prison gate at twilight. When they reached the first sentry, one of them spoke in a forceful voice to the effect that it was intolerable that a colonel should be the object of whistling and hooting by a detestable prisoner. He concluded by saying he would discuss the insolence with the general. The sentry took this discussion as proof of the officers' identity and came to attention. The invective was continued as the escapees approached the second sentry, who asked a timid question, but opened the gates when the disguised colonel growled an answer. The third sentry thought the first two sentries had checked the officers' identity and password so said nothing, but stood at attention, saluted, and opened the gate. The last hurdle was the guard at the moat. He asked for their passes, but they irritably told him that it was the third time they'd been called upon to show their papers and brushed him aside. He allowed them to move on.[55]

An escapee must have the presence of mind to exploit the suggestibility of individuals in a critical situation. One escapee took refuge in what he had thought was the railroad station waiting room. To his consternation it was an enemy officers' mess. Instantly reacting to the situation, he boldly walked over to the electric light switch, took it apart, and then replaced it. No one paid any attention to him.[56]

Evasion Techniques

Once an escape has been effected, the escapee is faced with the problem of getting out of the area and reaching a friendly border. The safest and best solution is to make contact with the local underground escape-and-evasion network. One escapee kept asking questions about how the war effort was going; when he met with favorable responses, he then inquired about the local evasion network.

Underground resistance organizations have certain security precautions in order to establish the reliability of persons requesting assistance. One common procedure is to interview the requester several times. The cross-examining technique is designed to bring out any discrepancies in his story.[57]

In many cases, however, escapees have had to rely strictly on their own resources. When traversing a country and evading security forces, a person seeks aid only from individuals and isolated homes, never from groups. When crossing a guarded bridge, or checkpoint, he should mingle with a large group of people crowding through.[58]

FOOTNOTES

[1] See Capt. Franklin A. Hart, "Jungle Ambush," *Infantry*, LII, No. 2 (March-April 1962), p. 25. In army tactical doctrine ambushes are included under the category of raids. Included in this category is the assault, an open attack on troops, barracks, bivouacs, etc. However, the literature on this subject uses the word "raid" interchangeably for an "assault."

[2] Fred H. Barton, *Salient Operational Aspects of Paramilitary Warfare in Three Asian Areas*, ORO-T-228 (Chevy Chase, Md.: Operations Research Office, 1953), pp. 57 and 91-92.

[3] *Ibid.*, p. 133.

[4] Ernesto Guevara, *Guerrilla Warfare* (New York: Monthly Review Press, 1961), pp. 63-64.

[5] *The New York Times*, February 7, 1964.

[6] George K. Tanham, *Communist Revolutionary Warfare* (New York: Praeger, 1961), pp. 89-90.

[7] Barton, *Paramilitary Warfare*, p. 206.

[8] *Ibid.*, p. 94.

[9] Malcolm W. Browne, *The New Face of War* (New York: Bobbs-Merrill, 1965), p. 76.

[10] Barton, *Paramilitary Warfare*, p. 202.

[11] Tanham, *Communist Revolutionary Warfare*, p. 89.

[12] Rowland S. N. Mans, "The Ambush," *Marine Corps Gazette*, XLVII, No. 2 (February 1963), p. 41.

[13] Barton, *Paramilitary Warfare*, pp. 92 and 131-32.

[14] *Ibid.*, pp. 94-95.

[15] William R. Peers, "Guerrilla Operations in Northern Burma," *Military Review* (October 1964), pp. 86-98.

[16] Hart, "Jungle Ambush," p. 26.

[17] Guevara, *Guerrilla Warfare*, pp. 63-64.

[18] Barton, *Paramilitary Warfare*, p. 133.

[19] Tanham, *Communist Revolutionary Warfare*, p. 100-101.

[20] *Ibid.*, p. 90.

[21] Maj. Fred K. Cleary and Linton C. Beasley, "Ambush," *Infantry*, LIII, No. 4 (July-August 1963), p. 47.

[22] Capt. William C. Dukes, "Viet Cong Vehicular Ambush," *Infantry*, LIV, No. 4 (July-August 1964), pp. 24-27.

[23] Maj. J. O. Langtry, "Tactical Implications of the Human Factors in Warfare," *Military Review*, XXXVIII, No. 11 (February 1959), pp. 85-89.

[24] Barton, *Paramilitary Warfare*, p. 96.

[25] *Ibid.*, pp. 96-97 and 113.

[26] *The Washington Post*, January 18, 1964, p. A-8.

[27] *The New York Times*, January 18, 1964, p. 14.

[28] *The New York Times*, July 5, 1964.

[29] John Maffre, "Tight Security Ordered After Viet Cong Attack," *The Washington Post*, November 2, 1964, p. A-1; Dennis Bloodworth, "Two Paths to Victory Eyed by Viet Cong Officials," *The Washington Post*, December 3, 1964, p. A-21; and John Maffre, "U.S. Beefing Up Viet Air-Fields," *The Washington Post*, December 6, 1964, p. A-1.

[30] Barton, *Paramilitary Warfare*, p. 57.

[31] Ronald Seth, *The Undaunted: The Story of the Resistance in Western Europe* (New York: Philosophical Library, 1956), pp. 44-45.

[32] Lawrence Wolfe, "Shadow Armies That Fight for the Allies," *The New York Times Magazine* (April 11, 1943), p. 7.

[33] Lt. Jans Lillelund, "The Sabotage in Denmark," *Denmark During the Occupation*, ed. Borge Outze (Copenhagen: The Scandinavian Publishing Co., 1946), pp. 53-56.

[34] J. D. Sanderson, *Behind Enemy Lines* (New York: D. Van Nostrand, 1959), pp. 170-89 and Seth, *The Undaunted*, pp. 42-47.

[35] Interview with former commander in the Yugoslav resistance.

[36] See Ladislas Farago, *War of Wits* (New York: Funk and Wagnalls, 1954), p. 240; David Martin, *Ally Betrayed* (New York: Prentice-Hall, 1946), pp. 177–79; and G. de Benouville, *The Unknown Warriors* (New York: Simon and Schuster, 1949), p. 197.

[37] John A. Armstrong (ed.), *Soviet Partisans in World War II* (Madison, Wis.: University of Wisconsin Press, 1964), p. 219.

[38] George K. Tanham, "The Belgian Underground Movement 1940–1944" (unpublished Ph.D. dissertation, Stanford University, 1951), pp. 85 and 202.

[39] *Ibid.*, pp. 84–85.

[40] Feliks Gross, *The Seizure of Political Power in a Century of Revolutions* (New York: Philosophical Library, 1958), pp. 350–51.

[41] Farago, *War of Wits*, pp. 240–41.

[42] J. Edgar Hoover, *Masters of Deceit* (New York: Henry Holt, 1958), p. 281.

[43] Andrew R. Molnar, et al., *Undergrounds in Insurgent, Revolutionary and Resistance Warfare* (Washington, D.C.: Special Operations Research Office, 1963), p. 119.

[44] Gunther Nollau, *International Communism and World Revolution* (New York: Praeger, 1961), p. 164.

[45] Peter Grose, "Vietcong Intensifying Efforts to Spread Disorder in Saigon," *The New York Times*, December 2, 1964.

[46] U.S., Department of State, *Aggression from the North: The Record of North Vietnam's Campaign To Conquer South Vietnam* (Washington, D.C.: Department of State, 1965), p. 5; see also Seymour Topping, "Portrait of Life with the Viet Cong—A Defector's Own Story," *The New York Times*, May 23, 1965, p. E–3.

[47] Gerry Robichaud, "Cuba Travel Spotlighted by Mexico," *The Washington Post*, March 27, 1963, p. A–14.

[48] Richard Collier, *Ten Thousand Eyes* (New York: E. P. Dutton, 1958), pp. 195–96.

[49] P. R. Reid, *Men of Colditz* (New York: J. B. Lippincott, 1954), pp. 15 and 18.

[50] David James, *Escaper's Progress* (New York: W. W. Norton, 1955), pp. 67–68.

[51] Reid, *Men of Colditz*, pp. 115–123.

[52] Aidian Crawley, *Escape from Germany* (New York: Simon & Schuster, 1956), p. 66.

[53] James, *Escaper's Progress*, pp. 93ff.

[54] *Ibid.*, p. 57.

[55] Lt. Anselme Marchal, "Hoodwinking the German," *Escape*, ed. Capt. H. C. Armstrong (New York: Robert M. McBride and Co., 1935), pp. 7–9.

[56] James, *Escaper's Progress*, p. 56.

[57] David G. Prosser, *Journey Underground* (New York: E. P. Dutton, 1945), pp. 95–99.

[58] See John Dunbar, *Escape through the Pyrenees* (New York: W. W. Norton, 1955).

PART VI

GOVERNMENT COUNTERMEASURES

INTRODUCTION

Governments have successfully employed a number of strategies and tactics to counter underground movements. Communist governments have usually resorted to immediate application of force to eliminate underground activity as soon as it appears. Poland, North Vietnam, and Tibet are examples. When coercive force is maintained, resistance eventually diminishes and the people become compliant and accept the existing situation. To date there has not been one successful insurgency against a Communist government.

However, countries with representative or constitutional forms of government are restrained in their response by moral, legal, or social considerations. Usually, an attempt is made to win the people to the active support of the government through social, economic, and political reforms. But all too frequently a government does not detect the underground's subversive activities in time. As positive programs fail, either because of the advanced stage of the underground movement or because of inadequate resources or time, a government must organize for more direct and forceful countermeasures involving police and military actions. This has been notably successful in Greece, Malaya, the Philippines, and Kenya. In Algeria, Palestine, and Indochina, it has failed.

Counterinsurgency operations generally commence with separate civilian and military commands, each of which has separate lines of authority. A unified command is then organized in order that all activities may be coordinated under a single commander. This was the pattern in Palestine, Malaya, the Philippines, and Algeria. Even when centralized direction of countermeasures is necessary, each area commander retains a certain amount of tactical autonomy. Because local conditions vary widely and changes inevitably occur, administrative control is kept flexible. Area commanders are frequently authorized and encouraged to independently effect civic improvements, pay informers, and take on-the-spot action to adjust deficiencies in either military or civic action programs.

Although the entire governmental apparatus is involved in the counterinsurgency effort, when the situation reaches a certain critical level the major responsibility rests on internal security forces—the civilian police and the military and paramilitary forces. The primary burden lies with the civilian police, who are charged with responsibility for maintenance of public safety and law and order. The armed services become involved when the insurgency reaches the militarization phase. Traditionally, the national army serves as a reserve internal security force. It has an inherent and implicit function as a psychological deterrent to the use of internal violence against the government, and military forces usually are assigned at least a part in maintaining law and order when the threat to the government is critical. In Malaya during the Emergency the members of the armed services were empowered to function in the same manner as civilian policemen in the interest of public safety.[*]

A critical feature of nearly all insurgencies, and a major problem in countermeasures, is that the movement's subversive activities and physical structure remain largely invisible—that is, underground—throughout the early phases of its organization. The appearance of guerrillas or paramilitary units are relatively late manifestations, as they are organized only when a broad underground base of leadership and support has been established.

Typically, the process of building a revolutionary base is characterized by various stages of recruitment of cadres, organization of clandestine cells, penetration of mass organizations, and acquisition and storage of military supplies. In conjunction, the underground usually launches a psychological offensive of agitation and propaganda designed to discredit the government and intensify and channel existing population discontent. In proper perspective, then, the first ambush of a government convoy or attack by a band of guerrillas is predicated on years of quiet, invisible preparation and organization.

Hence, in planning and conducting counterinsurgent actions, it is extremely important to recognize that although a guerrilla war is not evident, the insurgency may be well under way. Countermeasures which fail to take cognizance of the underground and the large amount of preparation necessary for an insurgent movement but focus only upon the guerrilla activity will be ineffective. Eliminating only the guerrillas leaves untouched the roots of the enemy's revolutionary structure. Counterinsurgency measures must be designed to simultaneously attack the entire structure—underground cells and leadership as well as the military arm—if more than a temporary and partial victory is to be achieved. It is basic to the planning and implementation of countermeasures against undergrounds that the vulnerabilities in their organizational structure and operational methods be identified.

ORGANIZATIONAL VULNERABILITIES

Operating clandestinely, undergrounds necessarily emphasize security. In organizing to meet their security requirements, they are implicitly committed to certain vulnerabilities in organizational structure. For example, undergrounds with a high degree of compartmentalization have little, if any, horizontal communication between cells. All written communications go through a slow system of cut-outs and intermediaries. Although this system is generally effective in providing some security for clandestine organizations, it has disadvantages. A defector can be reason-

---

[*] Federation of Malaya, *The Emergency Regulation Ordinance, 1948, with Amendments Made up to 31st March 1953* (Kuala Lumpur: Government Press), p. 27.

ably sure there will be no reprisal against him if all of the other members of his particular cell (who alone have knowledge of his membership) are eliminated by the security forces. This can be used to encourage defectors to inform. If the government can convince an underground that it has been infiltrated, the underground will normally take increased security precautions. The more severe the security restrictions, the more passive the underground and the less likely it is to perform dangerous overt activities against the government.

The organizational structure for intraunderground communications is also a potential vulnerability. Most underground communications are accomplished through the use of couriers and mail-drops. When they are discovered by security forces, they are usually not immediately apprehended but placed under surveillance so that other contacts can be identified. Continuing surveillance also often leads to the exposure of key functionaries and organizers of the underground. Where time and distance are factors, undergrounds may communicate through rapid systems such as the double-language technique in which apparently simple propaganda radio broadcasts contain messages and instructions coded in key words and phrases. Such broadcasts can be interpreted by security police to supplement other intelligence information.

Undergrounds often rely on mass organizations for legal facades and support. It is possible for a relatively small number of people to capture control of a mass organization through infiltration and manipulation of elections, but security forces who are aware of such techniques can alert organizations accordingly. Once police recognize that all members of an infiltrated organization are not necessarily members of the underground, actions can be taken to apprehend the cell members without antagonizing or alienating the innocent.

Underground administrative activities such as finance, training, and supply are normally centralized and often located in a nearby country. Governments may attack such international activities by cooperating, wherever possible, with neighboring countries in planning effective countermeasures or setting up points of surveillance in order to identify the underground agents and follow them to internal supply caches.

Underground organizations have an advantage over police in that they need not be restricted by administrative or territorial jurisdictions. Underground operations are often planned to take advantage of organizational interface problems of communication and coordination among government jurisdictions. Careful countermeasure planning can reduce this advantage by insuring that counterinsurgent forces are flexible and cooperate with each other at all levels. Government forces must reduce jurisdictional differences to a minimum and provide for local initiative.

OPERATIONAL VULNERABILITIES

The effect of underground revolutionary activity may be minimized if government police exploit the operational vulnerabilities in underground activity. Most undergrounds, for example, resort at some stage to mass recruitment, which increases the underground's vulnerability to infiltration by government agents. Also, undergrounds are a "normative-coercive" movement, and not all its members are voluntarily associated with it; indeed, many would leave or defect if provided a safe alternative by the government. If the personal and situational pressures that the underground member faces are understood, government appeals may effect high-level defections.

Vulnerabilities in underground propaganda and agitation operations depend on local conditions which determine whether the government should answer the underground's propaganda or remain silent. The British in Malaya, for example, attempted to distinguish between legitimate and fabricated grievances. Realizing that labor unions had been infiltrated by the Communist underground, the British imposed the restriction that they could deal only with legitimate labor problems. The Communists were robbed of a propaganda and agitation platform, but legitimate labor grievances were not overlooked.

Underground terror operations may be countered by either establishing effective and visible demonstrations of the government power to protect the populace from terroristic threats or by organizing mass neighborhood or worker groups to provide group protection and act as informers.

Government forces can minimize the extent to which an underground organization can manipulate a crowd if they are able to identify and suppress essential preriot preparations, and photograph or otherwise identify the agitators in the crowd.

Government countermeasures against underground-sponsored shadow governments must focus on the coercive structures which support them. The hidden elements of control—such as the clandestine cells for surveillance and terrorism—as well as the visible institutional coercive structures, such as courts, group organizations, and tax collections, must be destroyed. If the underground shadow government basically rests upon coercive sanctions, it is particularly vulnerable to countermeasures which substantially reduce their threatening force.

There have been as many counterinsurgency programs as there have been insurgent movements, but no one formula has emerged as being totally successful. All programs have certain common patterns of organization and operation, however. These elements of intelligence, population control, defection programs, and civic action are discussed in the following pages.

CHAPTER 15

COUNTERINSURGENCY INTELLIGENCE

TYPES OF SYSTEMS

The importance of intelligence to all counterinsurgent operations is unquestioned, but there have been several approaches used in developing intelligence organizations. A common type of intelligence organization is the unified system, where all collection and processing of intelligence are coordinated within one group under one command. In the Malayan insurgency, the intelligence efforts of the British Armed Forces and Malay security police were coordinated under the command of the British High Commissioner. All data and information were sent directly to one command for processing. Similarly in Algeria, French Army intelligence staffs performed all intelligence functions under a unified collection command.

The advantages commonly cited for the unified collection system are that intelligence information can be processed more rapidly, there is no duplication of effort, and fewer agents and staff personnel are required. However, a single channel for communicating intelligence is more vulnerable to compromise by underground infiltration, and in the unified system there is no independent source for confirming or cross-checking intelligence information and estimates.

The multiple organization of intelligence divides responsibility for intelligence collection and assessment functions among the various branches of government—the armed services, the civilian police, and the security police. For example, during the counterinsurgency operation against the Huks in the Philippines, intelligence functions were divided between the Military Intelligence Services (MIS) of the Armed Forces of the Philippines, the Philippine Constabulary, the National Bureau of Investigation, secret agents of the Philippine president, and intelligence agents of other countries. When Defense Minister Magsaysay initiated his program against the Huks, intelligence functions in the field were made the responsibility of the MIS, which assigned permanent intelligence teams to the Battalion Combat Teams; other agents operated out of MIS headquarters to augment a team or to work independent of the military commander in the area.

The multiple intelligence system has the advantage of providing independent cross-checks on intelligence and on the reliability of information, and it is less vulnerable to compromise. The diverse number of agencies serves to stimulate competition in the collection and dissemination of intelligence, which in turn improves the quality of output.

Coordination of the various arms of an intelligence organization is an essential prerequisite for the efficient and meaningful collection of information. The problem is, of course, always more difficult in a multiple system of intelligence organization. To coordinate their army and police efforts, the British in Malaya adopted a "war council" consisting of the chief officials of the civilian administration, the police, army, and air force, and headed by a director of operations. The day-to-day planning and coordinating sessions among the police, military, and civil authorities did much to bring about concerted political and military action. In order to develop a countrywide intelligence system, a map of the Malayan Federation was drawn up which showed the areas in which the terrorists were operating, the distribution of Chinese and Malays, and the areas in which squatters were living. The map, known as the "Briggs Map," was posted with information from every possible source and also showed specific terrorist targets such as lumber camps, rubber estates, and mines. Drop zones for aircraft were indicated. Resettlement sites were chosen on the basis of the information gathered.[1]

There are a number of advantages in assigning intelligence responsibilities to local police instead of solely to military personnel. Usually the police possess more information about complex factors in a local situation since they are permanently located in the area, as compared to military forces who come into the area on a patrol mission, make a few contacts, and then leave. The police are also better able to collect local political intelligence.

Further, there are some psychological dangers in using soldiers as substitutes for policemen. Insurgents may gain prestige by broadcasting the fact that the government finds it necessary to employ its strongest armed force against them. Soldiers are an alien force in the daily life of most people and their appearance lends itself to unfavorable propaganda by the underground.

In Malaya, for example, the British authorities believed that the average villager would be more inclined to entrust information to a policeman he had known all his life than to a strange soldier. On the other hand, a villager might have personal reasons for distrusting a policeman or might be impressed with the army's efficiency and ability to protect him. Under such circumstances, a villager would be more inclined to give information to the army.[2]

Hence, the balance sheet is mixed: although the police are a small organization, they nonetheless seem to be able to collect quality intelligence at the local level; on the other hand, the size of a military force's organization makes it more useful for mass intelligence collection.

CONCEPTS OF COUNTERINSURGENCY INTELLIGENCE

Intelligence organization and collection during an insurgency place new demands on conventional concepts of intelligence. Requirements for effective countermeasures add a new dimension to intelligence functioning.

In conventional war, a combat unit learns the location of the enemy from contacts between units on an established line of resistance, and intelligence is reduced to the standard technique of providing "essential elements of information."[3] In counterinsurgency, underground and guerrilla targets are elusive and transitory, and the life cycle and usefulness of intelligence are brief. A few hours determine the success or failure of an action. In short, rapid response to intelligence is of crucial importance in counterinsurgency.

In conventional warfare intelligence is not primarily concerned with individuals, whereas in counterinsurgency activities it focuses on individuals and their behavior patterns. The identity and whereabouts of the insurgents are usually unknown and their attacks are unpredictable. The underground lines of communication and the areas of underground logistical support are concealed from view. It is to these highly specific unknowns that counterinsurgency intelligence must address itself.

Another feature of counterinsurgency warfare that makes intelligence collecting difficult is that the underground operates autonomously or in small, compartmentalized units. Finally, because of the improvised nature of insurgent organization and the crudeness of its operations, counterinsurgency intelligence does not easily fit into standard categories.

LONG-RANGE INTELLIGENCE

The parameters within which a revolutionary movement operates must be considered. Frequently a centralized intelligence processing center is established to collect and coordinate the vast amount of information required to make long-range intelligence estimates.

Long-range intelligence focuses on the stable factors operative in an insurgent situation. For example, various demographic factors such as ethnic, racial, social, economic, and political characteristics of the area in which the underground movement takes place are useful in identifying the members of the underground. Since the activities of an underground are clandestine and covert, it is important to have descriptive data as well as predictive information in order to identify causes, groups, and individuals who are or who may become members.

Information about the underground organization, on all levels—national, district, and local—is fundamental in counterinsurgency, as is also data about underground recruitment, training, and supply. Other stable factors that must be considered include strategies and tactics of the underground; previous successes and failures of underground operations; data on the areas controlled by the underground; the characteristics of those who are recruited into as well as those who defect from the movement; and, finally, a thorough understanding of various environmental features such as terrain, borders, climate, and communication-transportation routes.

SHORT-RANGE INTELLIGENCE

Collection of specific short-range intelligence about the rapidly changing variables of a local situation is critical. Information on the identification of members of the underground, their movements, and their modus operandi must be gathered. Biographies of suspected underground members, containing photographs, detailed information on their places of residence, their families, education, work history, and associates, are important features of short-range intelligence. Usually this information is circulated at the sector-level command.[4]

During the Malayan insurgency, the British depended for short-range intelligence on the cooperation of villagers in furnishing information to local authorities. Ways of protecting informants from risk of underground terrorist counteraction had to be devised. In one approach, police visited each house in a village and gave its inhabitants a sheet of paper on which they were to anonymously write any information about underground activity in their village. The following day the police returned with a sealed ballot box and collected all papers, blank or not. Since every house was visited, the underground had no way of knowing who the actual informants were.[5]

In another technique, used in the Philippines, government forces leaked information to a village that a raid was to be made against the Huks. Such news generally prompted the insurgents in the village to leave. No raid would be made; the soldiers simply took group pictures of everyone in the village. On a subsequent, unannounced, visit to the village, the soldiers could identify those who did not appear in the earlier picture as possible Huk members and question them.[6]

Occasionally, parties or festivals were sponsored for village children at their school. While they were occupied with the festivities, an intelligence agent would pass among them asking innocent questions about their parents, their father's name, family birthdays, and if there were anticipated births, as well as other related social information. This revealed a great deal of information about Huk members; birthdays and expected births gave intelligence officers an indication of when certain Huk members might return home and be vulnerable to arrest.

To get intelligence about the method of underground communication between agents in the villages and forces in the jungles, Philippine army intelligence agents moved among the villagers disguised as fishermen, peddlers, or deliverymen to keep watch on the movement of suspicious persons.[7]

Search-and-cordon operations are also a common way of obtaining short-range intelligence. To protect the identity of informants, the British in Palestine interrogated everyone in an area separately in a private booth. Cordon operations are also useful in identifying outside elements within a village or town.[8]

CONTACT INTELLIGENCE

In conventional warfare, contact with the enemy is usually well established through frontlines or patrol action. In counterinsurgency, the problem is to identify and then locate the enemy. As frequently stated, in an insurgency "the front is everywhere." Even after members of the underground have been identified and their pattern of operations established, they must be located before they can be captured. Once a body of background knowledge has been developed, there are essentially four methods of obtaining contact intelligence.

Patrols

When some knowledge of the underground or guerrilla behavioral patterns has been developed from study of their past movements, patrols or police squads can search for physical evidence such as tracks or campsites. If there is a consistent pattern, patrols can be selectively dispatched on the basis of anticipated movements of the insurgents.

Low-Level Informants

All tips and leads, no matter how unreliable, are sought after on the assumption that the information may be helpful to cross-check or compare with other background information. Leads which are obviously false are eliminated, and those which are probable are followed up. Eventually the bits and pieces give a composite picture of the individual or cell and its pattern of behavior.

Forced Contact

Through the strategic hamlet program in Malaya, the British were able to force contact between the underground and the guerrilla. When people along the jungle fringes were relocated in hamlets, the guerrillas were cut off from their underground supply. This forced the guerrillas into the open to contact their underground support arm, the Min Yuen. The British set up observation posts in an area that showed increased activity in supplying food to the terrorists. Thus, farmers who were members of the underground were identified. Security forces then arrested some of these farmers, purposely ignoring other members of the underground. The insurgents were forced to go more frequently to the remaining farmers and obtain larger quantities of food. Then, when the police had identified each insurgent, they set up an ambush on the route normally followed.[9]

Informants

The use of informants is one of the most reliable and rapid means of obtaining the specific data required in contact intelligence. In Malaya the British commonly placed informants within important villages. Through a process designed to protect their identity, informants were able to pass information about the movement, position, and activity of insurgents almost immediately. This intelligence was received by the local security forces, whose commander was authorized to take immediate action [a] under his own authority with no requirement to seek approval from higher authorities.[10]

---

[a] Because contact intelligence is a highly perishable commodity, a 2- or 3-hour delay in response is critical. Hence, contact intelligence is not generally processed through normal intelligence organizations or procedures.

INFORMANTS

Informants from among the general populace can be induced to assist the security forces through various motives—civic-mindedness or patriotism, fear, to avoid punishment, out of gratitude, for revenge or jealousy, or for remuneration. The British counterinsurgency forces in Malaya offered large rewards for information leading to the capture of Communist underground members and guerrillas. In the Philippines, the security forces placed a bounty on captured Huks. The bounties ranged from $50 to $25,000; in 1961, a top Communist leader in the Philippines had a $50,000 price on his head.[11]

A major inducement to informants is the assurance of protection from reprisal. The anonymity of the informant must be maintained and the transfer of information from the source to the security agent must be concealed. Many techniques and devices have been employed to minimize the risk to the informant. The security agent and the informant may prearrange signals to coincide with everyday behavior. In the Philippines, the government forces used light aircraft to spot such signals—a certain arrangement of colors on a clothesline, windows or gates left open to a particular angle, a specified household item or farm tool placed in a certain spot in the yard.[12]

Informants also protect themselves by anonymously mailing information about insurgent activities directly to the police. In one such method, the informant tears a strip of paper from his letter. At the end of the emergency, he can present his strip of paper to the police; if it matches the torn part of the letter, he can receive his reward.[13]

AGENTS

In addition to regular government agents who attempt to infiltrate the underground organization, intelligence agencies often attempt to persuade or coerce captured members or prospective defectors into working for government forces while remaining ostensibly loyal to the movement. A corollary technique is to place a trusted agent in a critical job where he has access to classified information; this position makes him a prime target for recruitment overtures by the underground organization and he can subsequently serve as a double agent.

In recruiting agents the security police consider the various motives which lead people to serve in such a dangerous capacity. Money, adventure, and revenge are perhaps the primary motives. The security forces can, however, create another motivating factor by revealing to a person incriminating evidence of his low-level involvement with the illegal organization and offering him the alternatives of cooperating or facing public exposure and arrest.

In Malaya, for example, a rubber plantation worker was observed smuggling supplies to the Communist terrorists. The police let him continue until they had enough information on the operation and sufficient

238

incriminating evidence. Then, one evening the police stopped on a lonely road and exposed their evidence. The worker was faced with a dilemma: he could receive a 10-year jail sentence for aiding the terrorists, or he could be executed by the terrorists if they learned he had cooperated with the police. Because of clandestine security precautions, only five terrorists knew the worker's name. The police suggested that he could resolve his dilemma by giving them the names of these terrorists, thus collecting a reward and removing the threat of revenge at the same time.[14]

The underground usually places new, untried recruits in positions where they have no access to compromising information; only after thorough testing does a new member receive responsible posts. The government may be able to circumvent these security precautions. In an effort to find a suitable person for infiltrating the Huk organization, the Philippine military covertly contacted the relatives of several Huk commanders until they found a cousin who was willing to cooperate. After 2 months of special training, the cousin was sent to meet his Huk relative. As a relative he would have an entree, but to justify his desire to join the Huks, the Philippine military burned his house, imprisoned his brother, and evacuated his parents. The government of course, agreed to pay for all damages and inconvenience, but it could not inform his family until the project was completed. Because of these obvious grievances against the government, the cousin was accepted and made collector for the Huks' national finance committee (the underground supply arm of the organization). To enable the infiltrator to advance in the Huk ranks, the government helped him collect medicine, ammunition, and weapons. For 2 months the Huks received supplies from the government through the infiltrator. His ability as a collector and his relationship to a Huk commander won him a promotion and a new assignment as a bodyguard to Luis Taruc, the Huk leader. Through an elaborate system of signals and contacts, the agent was able to relay vital information to the security forces concerning the elite corps surrounding Taruc at his headquarters and information on members of the National Finance Committee.[15]

UNIT INFILTRATION

Government security force units and teams of varying size have been employed in infiltration operations against underground and guerrilla forces. They have been especially effective in obtaining information on underground security and communications systems, the nature and extent of civilian support and underground liaison, underground supply methods, and possible collusion between local government officials and the underground. Before such a unit can be properly trained and disguised, however, a great deal of information about the appearance, mannerisms, and security procedures of enemy units must be gathered. Most of this information comes from defectors or reindoctrinated prisoners. Defectors also make excellent instructors and guides for an infiltrating unit.

In Kenya (1954-55), the British used "pseudogangs" successfully against the Mau Mau terrorists. The pseudogangs were composed of ex-terrorists, loyalists, and, occasionally, disguised Europeans. Through instructions from the ex-terrorists the pseudogangs learned Mau Mau handshakes, oaths, and prayers. These teams, once thoroughly briefed and oriented to Mau Mau characteristics, went into the jungle and contacted the genuine Mau Mau gangs. The infiltrating team then either moved on after relaying information on the Mau Mau to regular security forces or took advantage of their position and captured the gang themselves.[16]

In the Philippines a volunteer force of 3 officers and 44 enlisted men was secretly trained to resemble a typical Huk squad down to the most minute detail. All items that would identify them with the army were removed from their persons and they were given items generally carried by the Huks: ill-kept weapons, reading material, indoctrination booklets, propaganda publications, and mementos from girl friends. Clothing taken from captured Huks was distributed to the men to insure against any giveaway in consistency or uniformity of dress. Captured Huk weapons were issued to these simulated Huks. As a final note of realism, two wounded enlisted men volunteered to join the force.

During their 4 weeks of intensive training, the trainees addressed themselves as brothers and comrades and sang Huk songs. Reindoctrinated ex-Huks who had been captured in the area where the infiltration was to take place served as instructors and around-the-clock critics of the methods and mannerisms of the simulated guerrillas.

While the unit was in training, another disguised army unit made a reconnaissance tour through the area selected for infiltration. This reconnaissance patrol noted trails and defiles along the route and attitudes of inhabitants in order to incorporate them into the cover story.

The kickoff of the operation was a sham battle between the simulated Huks and a Philippine army unit. This battle set up the impression the simulated unit was driven from its own area into the Huk area selected for infiltration. The local Huk underground bought the story and for 4 days the simulated Huks fraternized with the local Huks. They learned the identity of the Huk leaders, their modus operandi, and the names of civilian government officials who were secretly collaborating with the Huks.

The infiltrating group had made one mistake, however; they had more ammunition than the regular Huk units, a consideration which had been overlooked in the preparation. When the infiltrators realized they were under suspicion, they acted at once, attacking and wiping out two guerrilla squadrons. As a followup to this operation, two companies of the battalion combat team moved into the village and spent 3 weeks screening the inhabitants and arresting members of the underground. This operation had serious effects on the Huks; thereafter when two Huk companies met, they were stringent in their requirements of identity.[17]

In employing a disguised team, the selected men should be trained, oriented, and disguised to look and act like an authentic underground or guerrilla unit. Defectors and reindoctrinated prisoners often make good instructors and guides. The unit should have a cover story and a reason for being in a certain area. Each man in the unit should be briefed on the cover story so that no contradictions will arise. Obviously, a high degree of secrecy is necessary in training and deploying such units. In addition to acquiring valuable intelligence information, the infiltrating units can demoralize the insurgents to the extent that they will be overly suspicious and distrustful of their own units.

CORDON AND SEARCH

The cordon technique is often used for gathering intelligence when the populace does not cooperate for fear of reprisal from the underground, or if informers have not been developed or are difficult to locate. In the most common type of cordon operation, security forces surround a specified area, seal off entrances and escape routes, and search the people and property within the area. Two factors work for the success of the cordon: the element of surprise and the anonymity it affords possible informants. Usually, once the area is sealed off, the people are removed from their houses which are searched. They are then taken one by one into an interrogation booth. If the surprise has been effective, there will have been no time for prearranged stories, and people will tend to be confused and uncertain of what to do. The identity of those who actually inform will be unknown to others, and chances of reprisal will thus be reduced. By careful interrogation of this large number of people, contradictions will be discovered and a meaningful pattern of information can be developed.

Sometimes the cordon can be employed successfully in large cities. For 6 weeks in the fall of 1947, a British battalion successfully cordoned and searched 10 areas in the heart of Jerusalem. The cordon troops appeared before daylight in separate columns and surrounded the selected area. Through perfect timing and control, the cordon was set up before the men of the area awakened and had a chance to leave. Over 70 wanted men and a number of weapons were captured, and the remaining underground terrorists in Jerusalem were forced to lie low or leave the city.[18]

One of the most successful cordons was carried out in southern Johore, Malaya, in November 1956. Before daylight, 10 villages were simultaneously surrounded. The cordon was a complete surprise to the villagers, and the police seized 278 suspects. In this one operation, all the Communist supporters whom the villagers feared were removed, and information from the villagers was more easily obtained by the police. The terrorists, deprived of their usual suppliers, were forced to turn to others who were willing to report their contacts. This operation became the wedge that led to the eventual ridding the terrorists from the whole of southern Johore.[19]

In Palestine, the British discovered that the Jewish underground, the Haganah, was acquiring detailed intelligence about British plans. The Jewish underground often had advance knowledge of British moves, presumably obtained from civilians employed at British bases. On other occasions Jewish communities sounded alarms that would bring hundreds of nearby villagers into a cordon area, defeating British attempts to identify strangers within the village. The British began sealing off bases where civilians worked until the cordon operations were completed, and made elaborate cover plans to conceal the purpose of various preparatory activities. Reconnaissance was usually not possible without tipping off the populace, so planning depended on maps and photographs of the village area. Written orders were kept to a minimum and usually distributed only hours before the mission; participating troops were alerted at midnight or later. No telecommunications were used in assembling troops for the cordon, since they might be monitored by Jewish underground agents. Finally, troops were assembled only under cover of darkness and the operation carried out just before dawn.

The success of these operations depended on speed, coupled with detailed coordination and specific instructions. Roadblocks were immediately established. An inner cordon of soldiers surrounded the area to be searched in order to seal it off and prevent escape. An outer cordon was placed at important points around the village to prevent interference from neighboring villages and to act as reserves. Special enclosed areas, or "cages," were established and suspected persons were brought there for interrogation. Search parties rounded up the other villagers and detained them in a separate area, while other troops searched for hidden arms. Screening teams checked identification cards and photos against lists of suspects. After these search operations, suspects were transported to permanent detention camps.[20]

There are obvious limitations on the use of cordons; for example, in jungle or mountainous terrain, cordon and search may not be successful because of the difficulty in sealing off an area. There are dense areas of jungle or underbrush where insurgents can hide and a soldier in a line of searchers cannot leave the line long enough to make a thorough search. In the jungle, insurgents can hear the approach of an advancing patrol and go into hiding. Cordon-and-search operations were not successful in areas of the Malayan jungle or in the mountains of Cyprus and Arabia.[21]

One of the principal limitations of cordoning operations is the difficulty of catching the underground or guerrillas by surprise. Local underground agents or sympathizers can warn of approaching patrols through pre-arranged, inconspicuous signals. In addition, large bodies of police or soldiers inevitably advertise their presence, giving the insurgents time to camouflage or conceal themselves. Another disadvantage of the cordon-and-search operation is that if security forces fail, the underground propaganda units can spread doubt among the people about the government's ability to enforce its security responsibilities. When people lose confidence in the government, intelligence sources dry up and one failure leads to another.

SURVEILLANCE

Surveillance, the covert observation of persons and places, is one of the principal methods of gaining and confirming intelligence information. Surveillance techniques naturally vary with the requirements of different situations; the basic procedures, however, include mechanical observation —such as wiretaps or concealed microphones—observations from fixed locations, and shadowing subjects.

In Jerusalem in 1947, a special unit of 12 army men was organized for surveillance operations. The surveillance unit placed known and suspected underground terrorists under continuous observation. The persons whom the suspects contacted were also identified and placed under observation. This continued until the pattern of underground organization was pieced together. One surveillance team was able to identify and arrest as many of the underground in one 6-week period as an entire army battalion could in the same period with cordoning operations.[22]

If an intelligence agent plans to employ shadowing, he should learn as much as possible about the person before undertaking the mission.[23] Particularly important is a description of the subject, his habits of dress, and his manner of walking. The description should stress how the person appears from behind since the surveillant will be observing the person from that angle. The agent must dress to blend with the environment, and must give the impression he is interested in local activities and not the person being shadowed.

The security agent should avoid nervousness and haste, and should have a cover story prepared in the event he encounters the subject. It is also important to be familiar with the transportation system and the pedestrian routes in the area. If the security agent loses his subject, this type of information may help him to pick up the trail again. Any number of agents can be employed on such operations; the use of more than one allows agents to alternate positions, thus minimizing the possibility of recognition by the subject.

The same tactics apply in shadowing by automobile. The number of automobiles and persons used in the surveillance task depend upon the difficulty of the tail. It is more effective to use two cars than one. The risk of detection can be greatly lessened by frequently changing position. There should be at least two persons in each car. If one car is doing the shadowing, it should stay about 100 yards behind the car being shadowed; if a second car is used, it should follow the same distance behind the first. If the shadowing is to take more than one day, a different vehicle should be used each day.

In surveillance of fixed places, a preliminary survey of the surrounding area should be made. The character of the neighborhood, the inhabitants, and the buildings should be observed. The observation point should be chosen after careful reconnaissance. Likely spots include a room in a nearby house or business establishment. A person may disguise himself as one who would normally be in the area—a street vendor, a building employee, or an artisan. An effort should be made to photograph visitors. Descriptive notes should be kept on the identity of persons and their times of arrival and departure. All movements of the security agent to and from the observation post should be made unobtrusively; if it is necessary to have confidants, they should be kept to a minimum.

Intelligence agents conducting surveillance have numerous opportunities to obtain information without raising suspicion. In urban insurgency, particularly, there are many sources of information. As an example, in a country that has developed industrially to the degree where there is mass consumption, credit-rating agencies are a possible source of information. Intelligence can be obtained from telegraph messages, telephone toll-call records, hotel registration cards, military personnel dossiers, and employment records.[24]

Intelligence information is frequently obtained from the trash output of homes and the wastepaper of business establishments. Arrangements can be made with contractors to get the trash, or the trash bags can be switched at night.

The contact is an important person in intelligence. Contacts can be established with clerks in barbershops, grocery stores, and drugstores, deliverymen (milkmen, mailmen, laundrymen, newspaper boys), and repairmen. Servants, friends, and neighbors are also a source of information.

In some instances more than one contact is maintained within an organization; this affords cross-checking of information. Multiple contacts insure the flow of information in the event that one is absent. Using contacts for information requires the establishment of proper security measures because each source has opportunities to desert to the other side.

RUSE

The ruse is another common technique in collecting intelligence. It may be described as a form of inquiry in which the security agent assumes an identity which helps him win the confidence or allay the suspicions of a potential informant.

The essential tactic of a ruse is to create a front which makes it legitimate (and therefore not suspect) to inquire about a person. Since most people are reluctant to inform on others, a ruse can be used to make it appear that the agent is seeking help rather than information. For example, most people, if asked to watch for suspects who just robbed the agent, are willing to comply in order to help a person needing assistance. Agents have used this ruse in numerous situations. For instance, to get information about a person from the registration card at a hotel, an agent can approach the hotel manager and report that he has received a bad check from a guest and wishes to compare the signature with that of the hotel registration card. Although such checking is not usually permitted, most hotel managers comply under these special circumstances.

Furthermore, the ruse gives potential informants some reason to avoid telling the suspect and alerting him to the fact that he is under surveillance. Were an agent to directly and openly seek information, the fact that an official was conducting an investigation would become known to the suspect and give him a chance to escape or take other measures to conceal his activities.

The ruse can so camouflage the nature of an agent's questioning that second parties may be induced to provide information or to observe suspect persons without either making them feel that they are spies or that they should warn the suspect.

Suggestion is one of the simplest forms of ruse. This was demonstrated in the famous case of the "North Pole" underground radio operation from Holland to London in World War II.[25] German Army counterintelligence in The Hague learned of the clandestine radio and, after months of patient search, located, raided, and captured the Dutch operator along with his radio and codes. Although this operator was a patriotic Dutch underground member, he succumbed to German threats that his punishment would be severe unless he cooperated in continuing to make his scheduled transmissions to London under German direction. He agreed, partly because he knew the Germans were unaware that he had been taught how to notify London if his operation were compromised. The agreed signal was for him to make certain mistakes in letters of the Morse code. However, London failed to catch the signal and assumed that the clandestine operation was progressing as planned. The Germans had the operator transmit messages requesting additional agents, as well as supplies and sabotage equipment. British authorities complied, and over a period of about 20 months more than 50 British and Dutch agents and literally tons of supplies (including brandy, coffee, and cigarettes) were parachuted into the hands of the waiting Germans.

To keep the operation running smoothly it was necessary to send to London a safe arrival message after each drop. The Germans knew that every agent had a special code signal for London; the problem was how to get this code before the agent learned he had been captured. To accomplish this, the Germans used suggestion. They dressed Dutch-speaking Germans in Dutch civilian clothes and sent them along with Dutch collaborators to welcome the agents and supplies in the drop zones. As soon as the agents landed they were hurried to a Dutch farmhouse where they were heartily welcomed by their supposed friends. They were told that they would have to wait there until it was safe to transport them to the headquarters of the Dutch underground. They were advised of the necessity of immediately informing London of their arrival and were requested to reveal the code message. The agents, relaxing after the tension of the night drop, would freely give this message to their "friends" and it was promptly transmitted to London. At that point, of course, the real nature of the reception committee was revealed and the agents were sent to prison for interrogation. To deflect suspicion from the radio operator, the Germans dropped hints during the interrogations that made the captured agents believe the real traitor was someone in the London establishment.[b]

Through this simple use of suggestion the Germans were able to obtain important code messages from a large number of highly trained agents. The agent who found himself in a highly unstructured situation was susceptible to subtle pressure.

INTERROGATION

The interrogation of agents, informers, suspects, and captured or surrendered members of the insurgent organization plays an important part in government countermeasures. The types of information sought by interrogation include identity of insurgents; location of contraband items such as arms, ammunition, radio transmitters, and printing equipment; plans and operations; and the organizational structure of underground groups.

In the process of interrogation,[27] the interrogator's personality must command respect from and dominate his subject. His attitude and performance must be professional. To inspire full confidence, the force of the interrogator's personality should be tempered by an understanding and sympathetic attitude. The person interrogated must feel confident that he is talking to someone who is concerned about his viewpoint and problems.

---

[b] In 1943 two of the captured Dutch agents escaped and reached Gibraltar via the escape-and-evasion net which operated through Holland, Belgium, and France. The Germans promptly sent a message stating that the two were collaborating with the Nazis, so that if they reached London anything they said would be viewed as false information. When the two loyal agents arrived in London, they were confined in a maximum-security area and treated as if they were double agents. When the British and exiled Dutch authorities finally realized that the escapees were telling the truth, the transmissions were terminated and the operation ceased.[26]

The villager who has been forced to cooperate with the underground will tell his story much more readily if he feels that the interrogator both understands his helplessness and seems inclined to take his plight into consideration.

The interrogator must be alert, able to analyze the material he collects, and quick to detect gaps or contradictions. Perseverance is required to complete a successful interrogation, and self-control must be exercised. Acting ability is helpful, for the interrogator must sometimes show anger or sternness to induce the interrogated to talk, while at other times he must feign the sympathy, kindness, and friendliness of a helpful adviser.

When conducting an interrogation, the interrogator establishes the atmosphere and dominates the interview. He must take care to have no distracting mannerisms. The interrogator and the interrogated should sit face-to-face, without intervening furniture, in as sparse a setting as possible. The interrogated should be deprived of every psychological advantage: the door should be at his back, there should be no window, the walls should be bare, and there should be no distractions for him to take refuge in. There should be no way for him to avoid the interrogator. It is the responsibility of the interrogator to create a mood conducive to a confession, and he must provide the emotional stimuli that will prompt the interrogated to tell what he knows.[28]

Techniques of Interrogation

There is abundant evidence that coercive practices have never been particularly effective in eliciting information.[29] An individual who is deeply committed to an underground organization is highly motivated to safeguard the information he possesses. Frequently, however, individuals who say "go ahead and shoot" in the face of threat of death later reveal the same information under seemingly mild pressure.

Interrogations during insurgencies take place in a variety of environments ranging from the traditional police or intelligence situation in urban areas to the interrogation of individuals under field conditions. Regardless of the situation, time is usually a critical factor. In many instances, questioning conducted as soon as possible after contact deprives the interrogated of certain psychological advantages and capitalizes on his anxiety before he can prepare adequate psychological defenses.

There are many ways of gaining information from a resistant deceptive person, depending upon the ingenuity of the interrogator. Some of the techniques practiced by experienced interrogators include emotional appeals, pretending to have physical evidence or other incriminating information, and demonstration of sympathy for the interrogated.

Where two or more individuals are being interrogated, the one who appears to be most inclined to talk can be told that another has already given the pertinent information. He may then be persuaded to talk since he now has no reason to withhold information. If two persons known to have vital knowledge are each permitted to give statements containing deceptive information, the discrepancies may furnish a means of getting additional information. In another commonly used technique, two interrogators may work with the same subject, with one assuming a stern attitude while the other seems permissive. In many instances the individual being interrogated over a period of time will give information to the permissive interrogator when he is alone with him.

The technique frequently followed in interrogating a large number of suspects in an area, such as a village or cordon, is to arrest and question them individually. The anonymity and safety found in a group can be used as an inducement to give information.

A variation on the above technique was employed by the French in Algeria: suspects would not be immediately interrogated, but would be kept waiting for long periods of time. When they were tired and apprehensive, as well as protected by the anonymity of a group, they were more inclined to talk.

In the Philippines a person was taken away from his home area to be interrogated. Removed from the local environment, individuals more readily offered information about people in the village. Occasionally, interrogators visited villagers on holidays while they were feasting and drinking and were thus prone to respond to seemingly harmless questions. Another practice followed in the Philippines was to divide prisoners to be interrogated into four categories:

1. Prisoners captured in combat. The first few minutes of interrogation are usually the most informative when dealing with persons just captured. The prisoners are usually confused and unable to develop cover stories.
2. Prisoners who surrendered because they feared reprisals by their comrades.
3. Prisoners who were serving jail sentences for serious crimes. These long-term prisoners were interrogated mainly to obtain information about their associates and contacts. The Huks were known to actively recruit people with prison records.
4. "Special prisoners." These were usually friends or close relatives of Huks or government informers who were placed in protective custody.[30]

It is important in interrogation to use a technique that guards against giving information to the subjects. Usually the interrogator employs an oblique approach or a ruse to avoid giving clues as to the real object of the interrogation.

Polygraph

The polygraph or lie detector measures human physiological responses to emotional or stressful stimuli and has been used to detect deception or knowledge of crime-related information. The polygraph does not indicate whether or not a subject is lying but only measures the physiological responses which are related to deception. It must be interpreted by an interrogator and a trained operator. The interrogator asks the subject a question pertaining to some critical event, usually phrased to obtain a "yes" or "no" answer. The subject's physiological reactions are recorded and compared with his normal response pattern to neutral questions. Measurements related to the autonomic nervous system are preferred, since the subject has little or no voluntary control of these responses. Although there are many devices used to measure deception, the term "polygraph" is usually reserved for that device which measures cardiovascular, respiratory, and galvanic skin responses.

The polygraph has been used extensively in criminal investigations and has been suggested for field interrogation in counterinsurgency operations. The use of the polygraph in the field is fraught with difficulties, however, because of the need for time and skilled operators.[c]

Even after assembling many skilled operators and consuming many hours interviewing thousands of persons, only one agent may be identified. For example, at the close of World War II, when German prisoners were tested by polygraph for political reliability, one interrogator was able to examine only eight POWs a day.[32]

There are other problems in the use of the polygraph. Since lie detection depends upon eliciting a characteristic emotional and physiological response to critical questions, those individuals who do not respond in a characteristic manner cannot be detected. Psychotics or neurotics, pathological liars, and persons who do not feel that they have violated any normative or moral behavior (such as children who do not yet have a concept of right and wrong) will not react characteristically.[33] Other individuals who respond to critical questions with the characteristic emotional patterns associated with deception are not necessarily lying; a variety of extraneous factors, such as anger or fear, may cause a deception pattern. Some persons do not have a stable response pattern, and, when nervous or upset, their responses to neutral questions are so variable as to make it difficult to determine if the response is related to the critical question.

There is only a limited amount of information available on the biological and physiological differences of people in various cultures. It is known that culture influences the physiological behavior of individuals, but precisely what influence culture has upon how deception affects physiology is yet unclear. In some Oriental cultures truth and falsehoods are not considered so black-and-white as in Western cultures, and there is a tendency to think in degrees of truth and falsity. Hence, cultural variables must be considered when interpreting the results obtained from the polygraph.

In general, the polygraph is accepted as an aid to investigation and not as a substitute for other investigative techniques. In counterinsurgency, the polygraph may aid security forces in locating caches of ammunition or supplies, screening individuals for underground activity, and may potentially be useful in acquiring contact intelligence.

---

[c] In South Vietnam some success has been recorded in using portable lie detectors in the field. In experimental tests on 10 government soldiers suspected by an American adviser of being disloyal, one proved to be an agent of the Viet Cong, another had once been a Viet Cong member, and a third had occasionally assisted the first two. The lie detectors used were spring-operated from batteries and small in size, one pocket-sized. More important, their operation was so simplified that the average soldier could use them after only a week's training.[31]

FOOTNOTES

[1] Harry Miller, *Menace in Malaya* (London: George G. Harrap Co., 1954), pp. 209–10.

[2] Major General H. T. Alexander, *Proceedings of the U.S. Army Operations Research Symposium*, Part I (Rock Island, Ill.: Office of Research and Development, Department of the Army, 1964), p. 174.

[3] See Irving Heymont, *Combat Intelligence in Modern Warfare* (Harrisburg, Pa.: The Stackpole Co., 1960), pp. 65–67.

[4] Richard Clutterbuck, "The SEP—Guerrilla Intelligence Source," *Military Review*, XLII (October 1962), pp. 13–22.

[5] Harry Miller, *Menace in Malaya*, pp. 209–10.

[6] Uldarico S. Baclagon, *Lessons From the Huk Campaign in the Philippines* (Manila: M. Colcon and Co., 1960), p. 31.

[7] *Ibid.*, p. 141.

[8] R. N. Anderson, "Search Operations in Palestine," *The Army Quarterly* (January 1948), pp. 201–208.

[9] Richard L. Clutterbuck, "Communist Defeat in Malaya: A Case Study," *Military Review*, XLIII (September 1963), pp. 63–78.

[10] Richard L. Clutterbuck, "The Cold War," *The Army Quarterly and Defence Journal*, LXXXI, No. 2 (January 1961), pp. 161–80.

[11] See Clutterbuck, "The SEP," p. 16; Medardo T. Justiniano, "Combat Intelligence," *Counter-Guerilla Operations in the Philippines, 1946–1953*, Seminar (Ft. Bragg, N.C.: 15 June 1961), p. 47.

[12] Col. Napoleon Valeriano, speech at the Counterinsurgency Officers Course, Special Warfare School, Ft. Bragg, N.C.: November 5, 1963.

[13] *The Christian Science Monitor* (Boston), February 26, 1965, pp. 1, 6.

[14] Richard L. Clutterbuck, "An Anti-Communist Agent in Malaya," *Marine Corps Gazette* (August 1964), pp. 32–35.

[15] Justiniano, "Combat Intelligence," pp. 44–46.

[16] See Frank Kitson, *Gangs and Counter-Gangs* (London: Barrie and Rockliff, 1960); also Clutterbuck, "The Cold War," p. 167.

[17] Napoleon D. Valeriano, "Military Operations," *Counter-Guerrilla Operations in the Philippines, 1946–1953*, Seminar (Ft. Bragg, N.C.: 15 June 1961), pp. 32–39.

[18] Clutterbuck, "The Cold War."

[19] *Ibid.*

[20] Anderson, "Search Operations In Palestine," pp. 201–208.

[21] Clutterbuck, "The Cold War."

[22] *Ibid.*, p. 167.

[23] See Charles E. O'Hara, *Fundamentals of Criminal Investigation* (Springfield, Ill.: 1956), pp. 160–76.

[24] William S. Fairfield and Charles Clift, "The Private Eyes," *The Reporter* (February 10, 1955), p. 20.

[25] H. J. Giskes, *London Calling North Pole* (London: William Kimber, 1953), pp. 74–86; and Pieter Dourlein, *Inside North Pole* (London: William Kimber, 1954), p. 111.

[26] *Ibid.*

[27] See O'Hara, *Criminal Investigation*, pp. 95–114, and Albert D. Bidderman and Herbert Zimmer (eds.), *The Manipulation of Human Behavior* (New York: John Wiley and Sons, 1961).

[28] For an interesting example of the effect of magic and superstition, see Frank Kitson, *Gangs and Counter-Gangs* (London: Barrie and Rockliff, 1960), pp. 62–75.

[29] Bidderman and Zimmer, *Human Behavior*, pp. 12 and 27.

[30] Justiniano, "*Combat Intelligence*," pp. 42–43.

[31] "Use of Lie-Detectors Planned by U.S. to Identify Viet Reds," *The Washington Post*, February 29, 1964.

[32] Robert L. Berry and Benjamin J. Malinowski, "Lie Detection," *The Military Police Journal*, XIII (August 1963), p. 20.

[33] Maurice Floch, "Limitations of the Lie-Detector," *Journal of Criminal Law, Criminology and Police Science*, XL (1950), pp. 651–53.

CHAPTER 16

DEFECTION PROGRAMS

Defection programs have played a vital and significant role in the successful outcomes of the Philippine and Malayan counterinsurgency efforts. The psychological impact of defection on the underground is significant, and, in addition, information derived from defectors can provide counterinsurgency forces with intelligence for both military and psychological operations.[1]

TARGETS

Many insurgents join the movement for specific, situational reasons, rather than for ideological and political beliefs. Many others are coerced into joining. Thus, even though factors which keep such people in the movement—loyalty to fellow soldiers, political indoctrination, threats of retaliation, and simple inertia—must be combatted, there is simple reason to believe that in any underground there are many potential targets who can be persuaded to defect.

FACTORS RELATED TO DEFECTION

Defection is related to certain long-range factors, such as the insurgents' estimate of the probable outcome of the insurgency. If they feel that there is little hope for an insurgent triumph, the tendency to defect is more widespread. Individuals who have families are more prone to defection; in many cases, family loyalty comes before political loyalty. Another factor is the length of service in the organization. New troops defect more frequently than do older troops, and the most critical time for defection is after 6 to 18 months of service. Other factors, such as being wounded or being a member of a minority group, are also related to defection.

Short-range factors, such as adverse weather conditions, casualties, and disagreements with superiors, are also related to defection, as are heavy losses during an encounter. Defections are most likely to occur in areas where the movement has suffered reverses. Lack of food and other hardships, along with the necessity of continually moving about in order to evade government forces, often raise the defection rate. Poor health contributes to individual decisions to defect.

At the point of decision, defectors are usually most concerned with the treatment they would receive if they defected. They seek reliable information and most report that they do not believe government broadcasts or propaganda messages until they have tested them against word-of-mouth information from the civilian populace. Persons with firsthand information are considered most credible.

No matter how widespread the information about amnesty and rehabilitation may be, the insurgent is still uncertain and doubtful about the intentions of the government forces. Will he be punished if he surrenders? Next, he is concerned with the dangers inherent in the act of defection itself. Will he be shot trying to leave the underground? Will the underground carry out its threats against him and his family? Uncertainty about the future and fear of retaliation are the chief inhibiting factors.

COMMUNICATING WITH POTENTIAL DEFECTORS

Opportunities to surrender must be presented in such a way that an individual has some degree of confidence in the offer. Defectors should be seen in public and can be sent in teams to various villages to tell the inhabitants why they left the movement, and news from the villagers will reassure potential defectors that the actual defectors are well-treated. Families and friends of known insurgents are especially good channels of communication because they relay information to their relatives through letters or meetings. It is important to assure defectors that they will not be punished. Pictures of government camps and photos of both early and recent defectors in groups should be publicized to convey the message that defectors are alive, well, and have not met with bad treatment.

Leaflets aimed at potential defectors should be small enough to be easily hidden. They may be printed in the form of safe-conduct passes and should contain both information on how to defect and reassurances of good reception by government troops as well as information for the government soldier on how to treat the defector. Air broadcasts have not been effective in convincing insurgents to defect, as they doubt the credibility of this source; appeals made through families, friends, and other civilian channels are more likely to be believed.

CONTENTS OF APPEALS

Broad, general themes designed to induce defection on ideological grounds have been ineffective; the potential defector is more interested in factual information. News of local tactical defeats and losses makes a greater impression than word of nationwide gains or losses. Thus, defection can be induced even when the insurgents are winning on a national basis, if the counterinsurgents take advantage of the psychological impact of local losses.

Divisive propaganda themes such as "soldiers make all the sacrifices while their leaders get preferential treatment and favors" help aggravate dissatisfaction. All themes should contain reassurances about the individual's safety and future legal status. Even when an insurgent wishes to defect, the mechanics of doing so can inhibit him. It is important to tell him how to escape, how to surrender, and where to surrender. Since defectors are afraid of being shot by soldiers, appeals should inform them of specified civilians to whom they can safely surrender, and describe the location of defection points.*

INFORMATION FROM DEFECTORS

As a routine part of defection programs, defectors should be systematically interrogated to determine the specific factors that led them to defect. Their means of escape and their reaction to appeals can provide information for campaigns in particular areas or regions. Systematic monthly records should be kept, so that trends can be observed. It is important to determine whether defectors were in political or military organizations, whether they were Communist or non-Communist, their level of command, and what geographic area and ethnic group they represent. This information will be useful in planning future efforts against the insurgents.

ORGANIZING DEFECTION PROGRAMS

In carrying out a program for defection, a concentrated effort should be made to coordinate psychological and other military operations. If the government tells the insurgents that defectors will be given fair treatment and government soldiers shoot a man who is trying to defect, this will deter other defectors. Therefore, combat troops must be indoctrinated and trained to perform their duties in a manner which will encourage insurgent defection. If, during the course of a battle, the insurgents find themselves confronted with superior weapons, such as artillery, aircraft, or napalm, or suffer heavy losses, potential defectors are likely to take some immediate action. Therefore, battles should be followed by psychological operations campaigns which are equally hard-hitting and which contain offers of amnesty to those who surrender.

RELOCATION CENTERS

In the Greek insurgency, many young people who defected were placed in jails with hardened Communists who re-recruited them into the insurgent movement. Others were placed in camps with inadequate facilities and programs and, becoming bored, went back to the insurgents. One solution to such problems is to place young defectors on parole in a distant city. Since the identity of a lower-level insurgent is not known outside of his immediate cell or unit, the young insurgents are protected from retaliation by relocation to areas distant from the ones in which they had operated.

---

* One defector in Vietnam reported that he felt he had no choice but to defect to a government-controlled town, for the countryside and the fortified hamlets were vulnerable to the Viet Cong. There were political agents in the villages, and the countryside was patrolled by small Viet Cong units who shot suspected deserters or collaborators.[2]

If a relocation center is established, it is important to insure it against insurgency attack, or the whole defection program will suffer. A primary fear of defectors is recapture. The underground try to infiltrate relocation camps to carry out acts of terrorism and coercion, and large centers are more vulnerable than small ones to this threat. One answer is to establish widely distributed small reception centers where defectors can be screened initially into three groups: bona fide defectors, dubious defectors, and potential infiltrators. The bona fide defectors can be relocated and re-integrated with a minimum of processing and indoctrination and the sus-pect defectors can be kept under surveillance.

Insurgents may try to flood the camps with sick insurgents or civilians who need medical aid, and facilities of the camp must be adequate to handle this. Food, clothes, and other facilities must be provided to insure minimum comfort for the camp inhabitants and to provide for any large influx of defectors.

It is important that defectors be treated with respect and not as prisoners of war. No stigma should be attached to their defection or their former activities, and the camp personnel should be trained in fair and appropriate treatment of them. Civilian visitors, especially relatives of insurgents, should be permitted to visit the camps to get a first-hand view of the treatment given to defectors. Attempts should be made to clarify the future legal status of defectors. To allay suspicion and fear, a brief gov-ernment indoctrination program should be set up to inform the defectors of the government's efforts on their behalf and to tell them what will happen to them in the future. It is important that the program be designed not to seek retribution but to induce others to defect.

Defectors are anxious to talk to anyone about their problems, plans, and worries, so camp personnel should be instructed to be sympathetic listeners. Recreational activities can also provide some emotional release. Lectures obviously intended to transform defectors into loyal citizens are usually neither effective nor necessary. The defectors are committed to the govern-ment by the mere fact that they are in the camp, and lectures on the vices of Communism are not required. A better approach to securing their loyalty, and one which provides emotional release for the defectors, is to permit defectors to participate in guided group discussions on their com-mon experiences and the reasons which led to their defection. This tends to reduce latent fears or doubts; listening to other defectors reinforces each person's reasons for defection and builds new loyalties to replace former loyalty to the insurgents. The enthusiasm generated by such ses-sions is reflected in their attitudes and behavior after they leave the pro-gram and can have a significant psychological impact upon the civilian community and, through them, on potential defectors.

Training programs to help develop skills have been effective in past counterinsurgency efforts. Since a lack of the necessary skills for earning a living in modern society often contributes to a man's decision to join an insurgency, an opportunity to acquire skills that will provide for his future financial security gives assurance that he will not rejoin the insurgents. Such programs are also an inducement for others to defect in order to benefit from them.

In summary, one of the most effective tools for undermining underground and guerrilla morale is a defection program which is aggressively carried out and thoroughly coordinated with counterinsurgency military operations. Many have joined the insurgents through coercion or for highly specific grievances and can be persuaded to defect if they are convinced that they will not be severely punished. The most effective way to communicate this fact to the insurgents is through well-functioning, fairly operated defection programs. It is as important to advertise the program to the populace as it is to attempt to reach the insurgents directly. Residence at the center should be brief and aimed at the rehabilitation of the defector.

FOOTNOTES

[1] For additional information on defection programs, see Alvin H. Scaff, *The Philippine Answer to Communism* (Stanford, Calif.: Stanford University Press, 1955); for data on the work of the EDCOR program, Lucian W. Pye, *Guerrilla Communism in Malaya* (Princeton, N.J.: Princeton University Press, 1956), especially chapter 14 on "The Process of Disaffection"; Andrew R. Molnar, *Considerations for a Counterinsurgency Defection Program* (Washington, D.C.: Special Operations Research Office, 1965); Andrew R. Molnar, *et al.*, *Undergrounds in Insurgent, Revolutionary and Resistance Warfare* (Washington, D.C.: Special Operations Research Office, 1963), pp. 185–91; William J. Pomeroy, *The Forest: A Personal Record of the Huk Guerrilla Struggle in the Philippines* (New York: International Publishers, 1963) for a personal description of feedback; and Alexander H. Leighton, *The Governing of Men* (Princeton, N.J.: Princeton University Press, 1945).

[2] Seymour Topping, "Portrait of Life with the Viet Cong: A Defector's Story," *The New York Times*, May 23, 1965, p. E–3.

CHAPTER 17

POPULATION CONTROL

A fundamental feature of insurgency is the competition between government and insurgency for the support and allegiance of the local population. If the underground expands its control over the population, inevitably the government's authority is reduced. If the government controls the population, the underground is deprived of its principal source of supply, intelligence, and refuge; in short, they are isolated and their survival capability is severely reduced.

Population control seeks to accomplish two different, but integrally related, countermeasure objectives: to restrict the movement of the insurgents and to separate them, both physically and psychologically, from the general population. Although government psychological operations attempt to achieve population control largely through persuasion and the natural predilection of people to conform to laws and normative rules, there is a coercive element in many population-control measures.

TECHNIQUES OF POPULATION CONTROL

A number of important techniques have been developed in population control: collective responsibility, resettlement and relocation programs, legal controls, registration requirements, and food controls. Various organizational patterns have also been established.

COLLECTIVE RESPONSIBILITY

A common technique is the institution of collective-responsibility measures, which holds a group or group representatives responsible for antigovernment acts such as collaboration or sabotage. The random nature of collective-responsibility reprisals not only discourages the population from supporting antigovernment activities themselves, but makes the citizenry oppose all underground activity, lest they be punished. Further, it makes the underground hesitant to commit antigovernment acts when the probability of harm to its own people is present.

The practice of collective responsibility as a means of population control appeared as early as 221 B.C. in the Chinese Empire. Called the *pao chia* system, this form of control evolved over several centuries as the population was divided into progressively larger groupings of families—tens, hundreds, and thousands. Through these groupings, government authority was extended down to the family, the basis of Chinese society. The groups were supervised by the Emperor's district magistrate through the headman of each family. More important, the members of each group were held mutually responsible for one another's actions and theoretically practiced mutual surveillance and denunciation. It was a criminal act to fail to report a crime against the state, and failure to do so automatically incurred punishment upon both the offender and his group.[1]

Another technique of collective responsibility was used during the American occupation of the Philippines in 1901. To stop Philippine insurgents from receiving support from various towns and villages in Batangas Province on Luzon Island, the U.S. military commander implemented a policy of "reconcentration" and retaliation. Around each town in which U.S. troops were stationed a boundary was set up within which was sufficient space for Filipinos living outside to move in and build their homes. These boundaries were then patrolled by U.S. troops, a curfew was established, and no male adults were allowed to leave except by special pass. Further, no food was to be taken outside the boundaries. Whenever any property, such as telegraph lines, was destroyed, native houses in the zone were burned in retaliation. A proclamation was issued throughout the province warning that if any American soldier or cooperating native was harmed, a captured insurgent would be chosen by lot and executed.[2]

Throughout nearly all of occupied Europe, the Germans in World War II took reprisals against the civilian populace for acts of sabotage. Hostages were frequently taken to insure good behavior. In Poland, for example, the German Army established a system of collective responsibility whereby a list of village leaders would be periodically posted in public with the threat that if any act of sabotage occurred in the village every individual whose name appeared would be summarily shot. This, of course, worked as a forceful deterrent since no member of the village underground wished harm to befall one of his relatives or friends. To circumvent this, the Polish underground brought in outside agents, who knew none of the listed villagers, for sabotage activities.[3]

RESETTLEMENT AND RELOCATION

A common population-control measure is resettlement or relocation. When successful, resettlement effectively seals off the insurgents from the populace and denies them material or intelligence support. Close surveillance in the resettlement projects also protects the populace from terrorist retaliation and coercion by underground groups.

One form of resettlement is detention or banishment. During World War II, Nazi occupation forces frequently employed this control technique, shipping scores of suspected underground agents or collaborators to concentration camps or forced-labor camps.[4] Elaborate resettlement programs were set up by the British in Kenya and Malaya. In Kenya, British Army forces relocated Mau Mau family clans to seal off the collection of food and intelligence by terrorists and to provide an opportunity for observation of contact agents. The famous resettlement program of General Briggs in Malaya brought together widely dispersed elements into specially relocated villages that could readily be observed and defended against Communist attacks.

Between 1950 and 1952, under the Briggs Plan, 400,000 people were resettled into 410 defended villages. The British cut off outside sanctuary of the terrorists, closing the Thai border and patrolling the sea, while through aerial observation they made it nearly impossible for the insurgents to grow food in the jungle. To combat Communist propaganda efforts to disrupt the massive resettlement program, the British persuaded settlers to move voluntarily by offering tangible benefits of health and school facilities, and improved living conditions.[5]

Algeria provides another example of relocation techniques. In April 1965, the French Army resettled a number of villages from zones along the Tunisian and Moroccan borders, which had been sealed off in an effort to cut underground supply lines. In addition, the French declared areas with heavy activity to be special security zones, and resettled their inhabitants into camps.[6]

LEGAL CONTROLS

In Malaya in 1953 the British High Commissioner promulgated emergency regulations that imposed a number of legal controls on the population. These regulations restricted the use of firearms and weapons and forbade possession of underground terrorist documents, association with people carrying weapons or acting in a manner prejudicial to public safety, and disseminating false information. Imposition of collective punishment on an area was legalized. British police were given authority to detain subjects for 2 years without trial, to search without warrant, and to deport and banish suspected subversives. Unique legal controls were devised to counter Communist infiltration of labor unions. Unions were required to register; if a union participated in political activities or demonstrations, it lost its registration certificate and was forcibly disbanded. Another regulation required a union officer to have 3 years of work experience in the craft or trade; this effectively eliminated most Communist infiltrators and protected long-time union officers.[7]

Other legal control measures include restriction and regulation of organizational or public meetings and licensing and censorship of printed publications and radio broadcasts. Controls are often set up on a country's borders to eliminate over-the-border sanctuaries. The issuance of scrip and frequent changes of currency prevent the accumulation of large sums of money to finance undergrounds or insurgent movements.[8]

In occupied areas, German authorities confiscated radios and set up public address systems to broadcast only the news the government wanted the people to hear.[9] Search-and-seizure laws have been promulgated in some areas to prevent the underground from obtaining weapons. The Philippine Government, as a control measure, instituted a large-scale program to buy up all weapons from the population.[10]

REGISTRATION

Government surveillance and control of the population can be exercised through various forms of registration—individual and group registration, transient control, and population census. The national registration program adopted by the British in Malaya required every person over 12 years of

age to be registered; identity cards were issued with a photograph and thumb print.[11]

In South Vietnam, identification cards, complete with photograph and identification details, are mandatory for all persons over 18 years of age. To obtain an ID card, an individual must appear in person at the district headquarters with two witnesses, a birth certificate, and a letter of verification from the village chief. There he must fill out a questionnaire which is forwarded to the district security police where it is checked against his files. A complete dossier is then prepared. However, a program that registers only persons over 18 misses one group—the 12 to 18 year olds—who are prime targets for underground recruitment as couriers and purchasing agents, as well as jungle fighters.

Loss or theft of cards is a problem in any registration system. A familiar underground tactic is the confiscation of identity cards of all villagers. A government countermeasure used in South Vietnam is to require the payment of a substantial fee for the replacement of lost or stolen cards. This promotes individual precautions against loss and creates resentment against the insurgents for confiscating the cards.[12]

Another common problem in the ID system is that of underground forgery. One countermeasure is periodic reissuance of cards on a different format. Another is affixing group, instead of individual, photographs on the cards. For instance, a family group is photographed and each of the family's individual cards bears this picture. In order to forge such cards, the underground is forced to gather a similar group together for ID photographs.

Another example of registration is found in the Chinese system of posting placards on houses, listing all the adult males of the household. When the Japanese occupied China during World War II they introduced a similar system, using wooden plaques for the names of residents, and organizing the populace into family groups, in accordance with the ancient custom.[13]

A similar technique of population control is seen in the South Vietnamese organization of Mutual-Aid Family Groups. Each family group receives a number from 1 through 8, depending upon how many houses are in the *lien gia* (five to eight households). This number is placed on a plaque outside the house along with the number of households within the *khom* (25 to 35 households). If, for example, the household number is 4 and that of the family 25, the plaque is inscribed 4/25. When a new house is added it becomes integrated into the family group within which it is located and receives the next number. The plaque also indicates the number of people in the house, their relationship to the head of the household, and, by use of circles, their sex and their education. Red circles indicate male adults, yellow circles represent females, and green represents children. If any resident is illiterate, his or her circle is only partially colored. In the villages, family declaration forms are registered under hamlet names, house number, and family group to facilitate the census, and a copy is forwarded to the district office. All of this information provides the police with a means for checking on the inhabitants, and can aid in detecting hiding insurgents. In the urban areas a similar system for the organization of the population was established in February 1960.[14]

Transient control is also an important facet of population control. Homes of peasants and villagers have always served as refuges or safe houses for transient guerrillas and underground members, as well as for stragglers and deserters from government forces. In China during World War II, households were forbidden to lodge strangers, and were required to report to village authorities the presence of suspicious persons. Further, an individual who wished to move from one village to another had to make application through a village population officer. Population movements in South Vietnam are controlled through use of exit visas issued by the district chief after a villager's request has been authorized by his village chief. This system has been a source of discontent for many villagers, who wish to go to other districts for seasonal employment. Similar emergency regulations issued by British authorities in Malaya in 1953 required an inhabitant to apply for a permit before he could travel from his home.

Censuses also are used to support population-control measures. In South Vietnam, for example, the yearly census includes three categories of data: one for all families, another for all males 18 years of age and over, and a third for all reservists. The census of families facilitates the organization of the Mutual-Aid Family Groups, while the data on reservists and male population are used by military agencies.[15]

FOOD CONTROL

Underground food supplies depend upon clandestine purchases, thefts, and collections from the populace. To exploit this vulnerability, governments usually institute control measures on the production and distribution of food. Tight food controls force the underground to spend an inordinate amount of time seeking food.

To separate guerrillas from their source of supply, villages are sealed off. Many food-control measures have been used. In Algeria, the French Army specified that the food reserves in any one place must not exceed enough for 30 days, so as to lessen the opportunity for the underground to pilfer supplies from a large inventory.[16] The British food regulations in Malaya provided for the establishment of a central distribution depot and rationing. The rice ration was closely supervised and limited to a week's supply. Villagers were forbidden to take any food out of a village. These controls were reinforced with the threat that the village's ration would be further reduced and a rigid curfew established if any terrorist was found to be obtaining food in the village. As a result, the British frequently received information from villagers on the identification of those in the village who attempted to supply the Communist underground. Smuggling food out of villages became the Communists' most vulnerable activity. As a last resort they cleared jungle areas to grow their own rice. However, these areas were easily spotted by aircraft and then eliminated by ambush.[17]

IMPLEMENTATION OF POPULATION-CONTROL MEASURES

SOUTH VIETNAM

Population-control measures in South Vietnam have been implemented through Mutual-Aid Family Groups, Village Self-Defense Corps, hamlet chiefs, and National Security Police. Mutual-Aid Family Groups were first organized in 1957 by the government, not only to counter Communist insurgents, but primarily to create a spirit of unity, mutual assistance, security, and achievement in reconstruction and social works. Another purpose was to promote an understanding of government policies and to carry out government orders concerning tax collection. In setting up the program, emphasis was put on obtaining politically reliable hamlet chiefs and on the propagandizing and training of group chiefs as well as individuals within the groups.

Village Self-Defense Corps have also been organized under the direction of village police officers. Applicants for membership must present birth certificates, police records, and a certificate of good character. Each is then given a security check by the National Security Police and undergoes a month's training at the provincial self-defense headquarters. His training covers weapons familiarity, basic military instruction, basic law enforcement, and political indoctrination. The corps patrols the area around its villages, aids in tax collection, escorts village officials, and generally protects the village.

The village hamlet chief also plays a role in population-control measures. Usually it is his responsibility to explain and carry out government policies, to carefully watch the activities of the people, and to maintain security and order.

One of the principal agencies for the implementation of population-control measures in South Vietnam is the National Security Police (formerly called the *Sûreté*). Maintaining offices at each level of government down to the district, it provides coherence, direction, and surveillance. At the district level three agents are assigned who, while primarily concerned with the acquisition of political information, also process requests for identification cards and clear prospective government employees.

In summary, it is largely at the district level that the security and population-control system of South Vietnam is administered—through the army, the district National Security Police, the hamlet and local village chiefs, and the district administration of the Mutual-Aid Family Groups.[18]

NORTH VIETNAM

As in other Communist countries, in North Vietnam a combined government-military-party directorate has been established to institute strict physical, ideological, and economic controls. The most important control organization in North Vietnam is the Lao Dong Party (Communist Party). It prevails over and permeates all organizations of state control. The organizational structure of the party parallels the entire system of government and is intertwined with it. Officials of the Lao Dong Party dominate all organs of the central government. At the village level the members of local branches of the party are also the leaders in local government. Party members are in charge of police and militia, as well as youth and women's groups. In the army, political commissars function down to the lowest unit level and are responsible for the political attitudes of their units.

The primary intent of this party duplication and permeation of official governmental organizations is to make the people feel the persistent presence of the party and to educate the masses in "correct thinking." The Lao Dong exercises great influence over both the official and unofficial organs of state control.[19]

In 1961 a People's Supreme Inspection Institute was set up to cope with problems of "destructive antirevolutionary elements" to protect the economy, and to secure law and order. An additional responsibility of this agency, and possibly its primary one, is the inspection of law enforcement in government offices, work camps, enterprises, farm camps, and cooperatives. The Inspection Institute is the North Vietnamese version of the Communist Chinese People's Supervisory Committee, the channel for the mutual surveillance and denunciation of waste, red tape, and corruption of negligent state officials by the people. In the performance of their duties, the Inspection Institute teams rely upon their investigative personnel and on "denunciation letters" which are used to investigate and scrutinize various government agencies. These denunciation letters provide the authorities with a massive network of informers.

In rural areas, the cooperatives are also used for the implementation of population-control measures the size of cooperatives, operating below the basic administrative unit of local government, depends upon such factors as population density, topography, and location. Generally, they range in size from between 150 and 200 families in the Delta region to 20 or 30 families in the mountain areas. The effort to establish cooperatives, and to extend them by relocating thousands of villagers from the Delta to mountain regions, has made little progress because of peasant opposition. The government has attempted to make the cooperatives more attractive by providing economic inducements and civic action programs. They have tried to provide better management, have agreed to set aside 5 percent of the cooperative land for private use, and have set wage rates based on work performance rather than on needs.

To facilitate control in the urban areas of North Vietnam, special protection committees have been established. In order to make the people more cognizant of government orders and to maintain public order and safety, in Hanoi alone 4,600 block chiefs and deputy chiefs, plus 3,000 committee members, have been appointed to properly control the urban population.

Militia and self-defense units have also been organized as a major means of population control in North Vietnam. As in Communist China, these units are organized into platoons, companies, and battalions; they are formed at all levels of government in both urban and rural areas, and extend into the base of the society.

The political mission of the military is to indoctrinate the civilian population and to aid in the establishment of cooperatives. Through its powerful political directorate, the army conducts indoctrination programs, especially for the individual soldier. The army fulfills its economic mission by helping with the planting and harvesting of crops, developing cooperatives, and cooperating with civilians in organizing public works projects.

The major civilian security control agency in North Vietnam is the Cong An—the secret police—who operate both overtly and covertly under the immediate direction of the Ministry of Public Security, and ultimately under the Lao Dong Party. Although little is known about the secret police organization, it is believed that, like its Communist Chinese counterpart, it operates at every level of government. There are several province-level sections, one of which is the political protection section that deals with identity cards and travel regulations.[20]

FOOTNOTES

[1] Floyd L. Singer, *Control of the Population in China and Vietnam: The Pao Chia System Past and Present* (China Lake, Calif.: U.S. Naval Ordnance Test Station, November 1964), p. 5.

[2] W. T. Sexton, *Soldiers in the Sun* (Harrisburg, Pa.: Military Service Publishing Co., 1939), p. 281.

[3] Interview with former member of the Polish underground.

[4] Andrew R. Molnar, *et al.*, *Undergrounds in Insurgent, Revolutionary and Resistance Warfare* (Washington, D.C.: Special Operations Research Office, 1963), p. 162.

[5] *Ibid.*, pp. 170–73.

[6] Paul A. Jureidini, *Case Studies in Insurgency and Revolutionary Warfare: Algeria, 1954–1962* (Washington, D.C.: Special Operations Research Office, 1963), p. 111.

[7] Federation of Malaya, *The Emergency Regulation Ordinance, 1948, with Amendments Made Up to 31st March 1953* (Kuala Lumpur: Government Press), pp. 3–11.

[8] Molnar, *Undergrounds*, pp. 170–71.

[9] *Ibid.*, p. 167.

[10] Col. Napoleon D. Valeriano, Speech at the Counterinsurgency Officers Course, Special Warfare School, Fort Bragg, N.C., 5 November 1963.

[11] Molnar, *Undergrounds*, pp. 259–62.

[12] United States Operations Mission, *National Identity Card Program—Vietnam* (Saigon: Public Safety Division, U.S. Operations Mission, 1963), p. 4.

[13] Singer, *Control of the Population*, pp. 32–33.

[14] *Ibid.*, p. 76.

[15] *Ibid.*, p. 74.

[16] Richard L. Clutterbuck, "The Cold War," *The Army Quarterly and Defence Journal*, LXXXI, No. 2 (January 1961), pp. 161–180.

[17] Molnar, *Undergrounds*, pp. 172 and 245–62; Richard L. Clutterbuck, "Communist Defeat in Malaya—A Case Study," *Military Review* (September 1963).

[18] *Ibid.*, p. 14.

[19] *Ibid.*, p. 13.

[20] *Ibid.*, p. 13.

CHAPTER 18

CIVIC ACTION

The general target for government civic action is that vast majority of the population which does not officially participate in an insurgency. An additional target is the large, nonpolitical portion of the insurgent movement itself which can be persuaded to support the government through offers of amnesty, rehabilitation, and opportunities for a better life.

Civic action programs take many forms. The governmental administrative apparatus and social services may be strengthened to aid victims of the insurgents, and inequities in the distribution of social services can be corrected through employment and welfare agencies. Public health programs and medical aid have frequently been used to win support among the population, as has the extension of educational opportunities. Price controls and rationing have been used to protect the population from inflation. In areas where farming is important, methods of improving agriculture and land distribution programs have been used to win support.

The coordination of military and civilian tactics among the civil populace is important because a lack of coordination can undermine a whole program. In carrying out publicity for civic action programs, mass communications alone are not sufficient. In an environment where the individual may be under stress and fear, face-to-face communications are more effective.

OBJECTIVES

There are several considerations in determining overall objectives of civic action projects. Programs designed to gain popular support and cooperation for the government will differ from those designed solely to prevent cooperation with the insurgents. The ability of civilian administrators to carry out the program without the help of the military will influence the scope of the project. There is the question of whether the program should concentrate on purely civilian projects such as schools or on projects which also aid the military, such as roads. Another consideration is whether or not to provide civic action in insurgent-controlled areas. One author says that political, social, and economic reforms, no matter how much they may be wanted by the populace, cannot be effective in such areas. In Algeria, for example, the FLN killed Arabs who took advantage of land reforms.[1] The government must be able to provide security before civic action can be effective.

PROGRAMS

In Algeria the French carried on a major pacification program. The Army had over 1,000 special service units in the rural areas. Each unit was headed by a French officer of company grade and staffed with an assistant and a secretary. The officers established residence and head-quarters in the midst of the Arab settlement or village and prepared to administer to the needs of the people. The main effort centered on im-proving living conditions. French Army volunteers organized and taught school classes, helped build houses, sanitary facilities, and water supplies, and demonstrated improved agricultural and health practices. They pro-vided medical services and took the ill and injured to hospitals for surgery. French women physicians and civilian employees taught Muslim women how to care for their babies. This program won the support of the populace in many areas.[2]

The French made certain that each soldier realized that psychological actions which he might undertake were no less important than military ones. The soldiers were to be the agents of pacification: they were indoc-trinated and kept informed by weekly sessions in which current develop-ments and plans were discussed and the successes and failures of each week's efforts were reviewed. Written and oral arguments to be used among the civilians were carefully prepared and kept up to date.

The essential elements in the program were 1) reassuring the popula-tion that everything possible was being done to bring peace and to protect them; 2) visiting people and showing them sympathy and help; and 3) re-specting their customs and traditions. The troops were briefed on local customs and officers often visited marketplaces to talk to the people. The French found the spoken word to be more effective than pamphlets or slogans, and deeds to be the best form of propaganda. Overly stringent demands were avoided and promises were carefully kept. Corrupt local officials and those of dubious morality were removed.[3]

In the Malayan counterinsurgency program, after the labor movement had been cleared of Communists, the strengthening of public morale and the winning of support by involving the people more directly in the struggle were undertaken. Malays who feared the Communists provided ample support. In order to get the commitment of the Chinese community, the authorities urged them to join legal organizations as a means of expressing indignation against Communist activities: a non-Communist, Malayan-Chinese association was established, as well as the Independence of Malaya Party and the Malay Labor Party. In addition, political figures were brought into direct working relationship with the British Armed Forces and the Malayan Security Police. In this way, the populace was made to feel that it shared formulation of government policy. This program allowed for day-to-day coordination between local civil and military authorities. Despite the obvious lessening of security, the government made more information available to the public; the government felt that the support thus gained was sufficient compensation for any reduction in military security.[4]

In the Philippines, the political, military, and economic reforms instituted by Ramon Magsaysay were instrumental in ending the Huk insurgency. In the 1951 election, government authorities stationed teachers as poll clerks, used ROTC cadets to guard polling places, and placed soldiers at ballot boxes to prevent intimidation of voters. These protective measures were announced in advance, and Magsaysay served notice that any officials who abused their responsibilities would answer in court. He urged all citizens to report directly to him and his personal staff any complaints that they might have.

Magsaysay reorganized the Armed Forces. Officers who were not performing their duties were removed, promotions were made on the basis of merit, and soldiers who were caught stealing were punished in the presence of the villagers. The determined effort to rid the Philippine Armed Forces of inefficient, unreliable officers resulted in the dismissal of personnel of all ranks, from the Army Chief of Staff to local battalion commanders who were slow in carrying out counterguerrilla operations. Magsaysay's success in ridding the armed services of undesirable elements increased public confidence in the national government. Magsaysay had the military perform civic and social welfare missions in addition to their military operations against the Huks. Each military unit was assigned to civil affairs officer who maintained liaison with the local barrio police officials in civilian home guard units.

Judicial reforms were also carried out. Before 1952 the Communists exploited the fact that small landowners rarely gained justice when abused by the large landowners; Magsaysay made provisions for the peasant to have the right to legal counsel at the government's expense if he so desired.

The economic reforms were the most spectacular. Initially, rural civic betterment activities were conducted under the army's psychological warfare group but later they were administered as civil affairs. At the battalion level, battalion commanders and their civil affairs officers met with barrio heads and other civilian leaders to plan the defense of the farmers in the fields and the barrio self-defense procedures. This led to further discussions of needs and to the initiation of army support measures. Army personnel escorted agricultural agents into rural areas to introduce newer agricultural techniques; troops were used to construct barrio schools, drill water wells, and carry out public works projects; civilians wounded in any battle between the government and the insurgents were treated at army hospitals.

The major civic action effort was undertaken by the Economic Development Corps (EDCOR). This project aimed at inducing defection from the Huks by rehabilitating and resettling Huk prisoners and their families. EDCOR formed four communities for former Huks, and established a vocational training center. Troops helped the settlers clear land and build houses. Village centers, school buildings, chapels, and dispensaries were set up. The army also assisted the settlers by handling the legal matters involved in land ownership.

Although the program was not large, the psychological effect of the EDCOR operations upon the Huk movement and the general populace was great. It provided new respect for the government and offered the rebels an alternative to resistance. Already assured of amnesty and protection from fellow insurgents, the rebels now had, through EDCOR, hope of future economic security. Huk defections eventually provided the government forces with the intelligence necessary to break the movement.[5]

PLANNING

In designing a civic action program, local conditions—political circumstances, cultural values, ideology, and technical and managerial skills—must all be considered. Those segments of society which would be most affected by the program should be consulted to determine the impact of the program. If interest groups are involved—trade unions, professional or women's groups, youth, peasants—they should also be contacted and, if possible, involved in the planning.

How should the program best be carried out? How will the people be informed and persuaded: leaflets? word-of-mouth? broadcasts? If people cannot be persuaded to voluntarily take part in the plan, what sanctions will be imposed: licensing? rationing? punitive measures? Is there a present crisis situation which predisposes the people to cooperate, or will it be necessary to create a crisis? Will the program be hindered by illiteracy, lack of social discipline, draft evasion, or underground sabotage? Will there be periodic review? Is the plan flexible? What side effects may occur? Will prices go up? Will attitudes change?

In Malaya squatters were relocated to villages in which they received schooling, housing, electrical facilities, and other benefits they had never had before. This was a step up the economic ladder for them, and many remained even after the Emergency was over. In Vietnam, on the other hand, individuals were awakened in the early hours of the morning by military troops and taken at bayonet point to their new location, while their houses were burned and their fields destroyed behind them. At the new hamlets they had to build their own houses and fortifications. They were often not compensated for the land and property left behind and even when they were, in many cases the new areas were not as desirable as the areas they had left.[6]

In Malaya, the villagers lived on large, easily enclosed plantations and did not have to leave their local areas. In South Vietnam, the fact that the population was scattered made it difficult to relocate people into strategic hamlets. Further, the value they placed on land left to them by their fathers made them reluctant to leave it.

It is important to anticipate complaints. People living under stress often react aggressively, civic action programs should capitalize on normal human tendencies to seek prestige, and redirect aggressions in this way.

Most civic action programs are more positive than coercive by nature, but most involve some element of force. In the Philippines, Magsaysay's reform program was called the "Iron Fist in the Velvet Glove": the people were given positive incentives to support the program, but there were penalties against those who did not. The program tried to make individual goals coincide with national goals and to offer some alternatives to resistance or apathy. Another characteristic of the program was the high degree of specific direction. Civic mobilization, and the use of the organizational incentive and similar controls, is one method of developing support for government goals.

In many civic action programs the government undercuts underground propaganda by offering many of the same things: for example, land reform or, in the colonies, independence. In addition, individuals not firmly committed to the insurgency were won to the government side by asking

their cooperation in efforts to correct grievances or inequities in the system. In Malaya, there were restrictions on direct criticism of the government, but union leaders were permitted to make demands or complaints about union matters. Another way used to overcome resistance is to legalize nonsubversive opposition. The creation of political parties and labor groups in Malaya was one way of accomplishing this.

If special areas are set up in which civic action programs can be tested before being used on a national basis, the effect of errors will be minimized and they can be corrected more easily. Also, critical side effects can be discovered by such testing. In Vietnam, for example, in order to combat inflation in the urban areas, price-control measures set the ceiling on pork so low that farmers refused to deliver their animals to Saigon; in thus attempting to satisfy one segment of society another segment was alienated.[7]

In addition to winning support from the civil populace, programs such as EDCOR and relocation camps can be attractive to the insurgents, and this should be recognized and exploited in planning civic action programs. Defection programs must consider the safety of the defectors. (In the Philippines and Malaya, defectors were relocated to areas which were pacified or within government control.) Another factor in inducing defection seems to be the importance of reeducation which offers a hope of a better position within the society.

COMMUNICATION

A basic part of civic action programs is redirecting the frustrations and aggressive feelings that people experience during the stress of insurgency into channels of action favorable to government purposes. To discover and manipulate such feelings requires communication, and communicating with people under stress involves some peculiar difficulties.

NEED FOR RELIABLE AUTHORITY

In a crisis or previously unexperienced situation, the scope of information that the individual is interested in is greatly reduced: he wants specific information bearing directly on the crisis. Information about vague, abstract future threats has a very low attention value, and people respond to it apathetically.

People want reliable information, and authoritative figures or emerging leaders can be highly effective in communicating in emergency situations. In addition, such people can assist the individual in making decisions.[8]

NEED FOR POSITIVE ACTION

Prolonged conditions of unresolved crisis lead to emotional depression; people feel a need to take positive action. In seeking a basis for action, the individual tends to rely upon the familiar, to integrate the crisis event into a past frame of reference. His tendency to act in familiar ways which have proven reliable in the past can lead him to take actions which seem illogical to others but which appear perfectly logical and appropriate to him. Government programs can take advantage of these tendencies, and overcome some of the difficulties they pose, by drilling people in actions to be taken during emergency situations. With such training, people can perform adequately; without it, they are highly receptive to suggestion.[9]

NEED TO COMMUNICATE

In changing or controlling attitudes during a crisis situation, free verbal expression serves to vent strong emotion and provides a catharsis. It gives the individual the chance to state his own position, opinions, fears, and desires.[10] The creation of some mechanism for this purpose is important and can take many forms.

CONSIDERATIONS FOR COUNTERMEASURE PROGRAMS

Since the insecurity and frustration generated by a crisis situation can be used by the insurgents to direct aggressive behavior against the government, it is important for the government to establish channels of communication with the people. The channels can provide the needed catharsis. The search for a reliable authority also lends itself to government purposes: the local government agent can fill this role and usefully spend his time listening to the hopes and fears of the people. The most effective way to alter the individual's perception of the situation is to direct him into constructive action, rather than simply to lecture him. Decisions made by a group carry more weight with the individual than those made independently; the individual who has shared in the decision of the group is more likely to go along with it.

In Malaya, the British had an unexpected bonus in their hamlet programs: local officials were bombarded with complaints and so began a dialogue between the villagers and the government which satisfied the former's psychological needs.[11] In the Philippines, Ramon Magsaysay made it know that anyone could send him a telegram for 5 cents complaining about anything and that within 24 hours his office would take action on the complaint.[12] In South Vietnam, in April 1964, Prime Minister Khanh ordered the establishment of a complaint agency to be called the General Office for People's Suggestions and Complaints. This agency was to investigate complaints and report the findings to the Prime Minister with appropriate recommendations. The director was given the authority to contact any military or civilian agency within the government and swiftly settle any claims. All other government agencies were required to help with suggestions and complaints.

Besides setting up protest mechanisms, it is important to have a feedback on how well a program is doing, so that rapid responses can be made to complaints, grievances, or inequities in the civic action program. The government's intent is most credibly conveyed through personal experience and word-of-mouth. In addition to information campaigns, the British in

Malaya sent defectors around to the various villages to lecture the people. Mayors from the various areas were brought to Singapore to see the reforms and actions the government had taken. Word of these projects filtered back to the villagers and had a greater credibility than anything that could have been said in a mass communication program.

FOOTNOTES

[1] David Galula, *Counterinsurgency Warfare* (New York: Praeger, 1964), p. 79.

[2] Andrew R. Molnar, *et al.*, *Undergrounds in Insurgent, Revolutionary and Resistance Warfare* (Washington, D.C.: Special Operations Research Office, 1963), pp. 280–82.

[3] Slavko N. Bjelajac, "Psy War: The Lessons From Algeria," *Military Review*, XLII, No. 12 (December 1962), pp. 2–7.

[4] Molnar, *Undergrounds*, pp. 259–62.

[5] *Ibid.*, pp. 325–29.

[6] Denis Warner, *The Last Confucian* (New York: Macmillan, 1963), pp. 17–18.

[7] Homer Bigart, "Saigon Is Losing Propaganda War," *The New York Times*, January 17, 1965.

[8] Harry B. Williams, "Some Functions of Communication in Crisis Behavior," *Human Organization*, XVI, No. 2 (Summer 1957), pp. 15–19.

[9] F. P. Kilpatrick, "Problems of Perception in Extreme Situations," *Human Organization*, XVI, No. 2 (Summer 1957), pp. 20–22; and Otto Klineberg, *Tensions Affecting International Understanding* (New York: Social Science Research Council, 1957), p. 162.

[10] Klineberg, *Tensions*, pp. 162 and 165.

[11] Ralph Sanders, "The Human Dimension of Insurgency," *Military Review*, XLIV, No. 4 (April 1964), pp. 46–47.

[12] Charles T. R. Bohannan, "Antiguerrilla Operations," *The Annals of the American Academy of Political and Social Science*, CCCXLI (May 1962), p. 26.

APPENDIX A

METHODOLOGICAL APPROACH

The study of motivation and behavior in controlled environments is always difficult. Analysis based upon data which is incomplete and often biased, and which was recorded in an uncontrolled, hostile environment, is even more difficult. When insurgencies are in process, it is impossible to determine the actual motives of the individuals involved. The best evidence available appears in the reasons given by the people concerned: why they did things, what they did, and what consequences ensued for them.

In analyzing the organization, motivation, and behavior of insurgents, the following sources of data were used:

1. *Historical accounts of underground movements.* The accounts of various insurgent underground movements were reviewed to establish the environmental context within which the events occurred.

2. *Organization charts.* Although organization charts are only formal statements of the organization, comparisons were made among various insurgent movements in an attempt to identify similarities and differences in structure and functions, command and control, and communications.

3. *Training materials.* Although it is difficult to establish exactly what went on at various insurgent training centers, a review of the materials used suggests the behavioral patterns instilled in the recruits.

4. *Propaganda.* It was possible to evaluate propaganda messages and isolate those commonalties which undergrounds agreed upon as appealing to human motivation.

5. *Interrogations.* Records of interrogations of defectors were reviewed to discover the reasons they gave for joining, remaining in, and leaving the movement.

· 6. *Autobiographies.* By using autobiographical material it was possible to identify critical incidents and individuals' interpretations of the insurgent situation.

7. *Interviews.* Interviews were conducted with ex-underground members, former insurgents, and counterinsurgents in an attempt to obtain information not available in written documents.

8. *Captured communications.* Captured documents and afteraction reports also provided insights into the motivation and behavior of the insurgents.

9. *Mission analyses.* An attempt was made to determine the objectives of the various missions, the steps that were taken to accomplish the objectives, what alternate solutions were used at various times, and the consequences of the actions.

Although each of these various sources of information is suspect in and of itself, consistency among the various approaches leads to some degree of confidence in the conclusions.

In evaluating the various accounts, events, and data, the following criteria were used:

1. Was it possible that the event could have occurred?

2. Was it credible? Was there agreement among several sources who reported the event?

3. Was it possible to establish the occurrence of similar events, behavior, or motivation in other movements in different parts of the world?

4. Was there internal consistency (compared with other missions and operations)?

APPENDIX B

ANALYSIS OF 24 INSURGENCIES

The following seven tables present background data on countries that have experienced insurgencies since World War II. Most of the insurgencies have occurred since 1946, and all but two have been major ones in terms of disruption of government control, number of persons involved, or length of time covered.

The intent has been to describe in gross terms certain background information about each country. Thus, countries are ranked in quartiles for comparison with other nations and the world average. The data were drawn from Bruce M. Russett, *et al., World Handbook of Political and Social Indicators,*[1] and is presented in table IX. Comparisons are made using this data in subsequent tables (IX through XV).

Table IX. Composite Background Information on Countr

Country	Dates of Insurgency	G N P Per Capita ($) 1957	QUARTILE RANK	G N P Per Capita % Annual Change	QUARTILE RANK
1. Algeria	1954–62	178	II	4.7	II
2. Angola	1961–	60	IV		
3. Burma	1948–60	57	IV	3.9	II
4. Cameroon	1955–62	105	III		
5. Colombia	1948–	263	II	2.2	III
6. Congo	1960–	92	III	2.8	III
7. Cuba	1953–59	431	II		
8. Cyprus	1954–58	467	II	1.4	IV
9. Greece	1946–49	340	II	5.3	II
10. Haiti	1958–64	105	III		
11. Hungary	1956	490	I	4.8	II
12. Indonesia	1946–49	131	II	1.5	IV
13.	1958–61				
14. Iraq	1961–64	156	II		
15. Israel (Palestine)	1945–48	726	I	5.8	I
16. Kenya	1952–60	87	III		
17. Laos	1959–62	50	IV		
18. Madagascar	1947–48	88	III		
19. Malaya	1948–60	356	II	0.6	IV
20. Philippines	1946–54	220	II	2.4	III
21. Vietnam (RVN)	1956–	76	III		
22. Vietnam (DRV)	1956	55	IV		
23. Yemen	1962–	50	IV		
24. Venezuela	1958–	648	I	4.4	II

Table X. Occurrence of Insurgency and Gross National Product Per Capita for 24 Insurgencies [3]

Quartile Rank based upon 122 countries	GNP per capita ($—1957)	No. of Insurgencies in Which			Total
		Insurgents won	Gov't Won	Currently (1965) Undecided	
I	479–2,900	1	1	1	3
II	106–478	4	5	1	10
III	71–105	1	3	2	6
IV	45–70	0	2	3	5
	Totals	6	11	7	24

% Urban (20,000)	QUARTILE RANK	% Adult Literacy	QUARTILE RANK	Higher Ed. per 100,000	QUARTILE RANK	% Military (ages 15–64)	QUARTILE RANK	Outcome of Insurgency
14.1	II	19.0	IV	70	III			Ins. Won
4.7	IV	2.5	IV					Unclear
10.0	IV	47.5	III	63	IV	0.49	III	Gov. Won
4.1	IV	7.0	IV					Gov. Won
22.4	II	62.0	II	296	II	0.27	IV	Unclear
9.1	IV	37.5	III	4	IV			Unclear
36.5	I	77.5	II	258	II	0.84	II	Ins. Won
13.6	II	60.5	II	78	III			Ins. Won
38.4	I	80.0	II	320	II	2.52	I	Gov. Won
5.1	IV	10.5	IV	29	IV	0.29	IV	Gov. Won
37.0	I	97.0	I	258	II	1.14	II	Gov. Won
9.1	IV	17.5	IV	62	IV	0.24	IV	Ins. Won
								Gov. Won
23.6	II	10.0	IV	173	III	2.02	I	Gov. Won
60.9	I	93.7	II	668	I	4.84	I	Ins. Won
3.8	IV	22.5	III	5	IV			Gov. Won
4.0	IV	17.5	IV	4	IV	2.52	I	Unclear
7.6	IV	33.5	III	21	IV			Ins. Won
22.7	II	38.4	III	475	I	1.37	II	Gov. Won
12.7	II	75.0	II	976	I	0.28	IV	Gov. Won
		17.5	IV	83	III	2.00	I	Unclear
						2.90	I	Gov. Won
0.1	IV	2.5	IV			0.73	II	Unclear
47.2	I	52.2	II	355	II	0.49	III	Unclear

*Table XI. Occurrence of Insurgency and Percentage of Annual Increase of
Gross National Product Per Capita for 13 Insurgencies* [4]

Quartile Rank Based Upon 68 Countries	% Annual Increase GNP per Capita	No. of Insurgencies Where			Totals
		Insurgents won	Gov't Won	Currently (1965) Undecided	
I	5.5–7.6	1	0	0	1
II	3.3–5.5	1	3	1	5
III	1.7–3.7	0	1	2	3
IV	−2.2–1.6	2	2	0	4
	Total	4	6	3	13

*Table XII. Occurrence of Insurgency and Percentage of Population in Urban Areas
for 22 Insurgencies* [5]

Quartile Rank Based Upon 120 Countries	% in Urban Areas (+20,000)	No. of Insurgencies in Which			Total
		Insurgents Won	Gov't Won	Currently (1965) Won	
I	33.2–81.9	2	2	1	5
II	12.1–33.1	2	3	1	6
III	10.1–12.0	0	0	0	0
IV	0.0–10.0	2	5	4	11
Total		6	10	6	22

*Table XIII. Occurrence of Insurgency and Percentage of Adult Literacy
for 22 Insurgencies* [6]

Quartile Rank Based Upon 118 Countries	% of Adult Literacy	No. of Insurgencies Where			Total
		Insurgents Won	Gov't Won	Currently (1965) Undecided	
I	93.8–95.8	0	1	0	1
II	47.6–93.7	3	2	2	7
III	19.4–47.5	1	3	1	5
IV	1.0–19.3	2	4	3	9
Total		6	10	6	22

*Table XIV. Occurrence of Insurgency and Number of Students Enrolled in
Higher Education for 20 Insurgencies* [7]

Quartile Rank Based Upon 105 Countries	Students Enrolled in Higher Education (per 100,000 population)	No. of Insurgencies Where			Total
		Insurgents Won	Gov't Won	Currently (1965) Undecided	
I	461–1,983	1	2	0	3
II	194–460	1	2	2	5
III	70–193	2	1	1	4
IV	3–69	2	4	2	8
Total		6	9	5	20

*Table XV. Military Personnel as Percentage of Population in Countries Where
Insurgencies Have Occurred* [8]

Quartile Rank Based Upon 88 Countries	Percentage of Military in Population (aged 15–64)	No. of Insurgencies in Which			Total
		Insurgents Won	Gov't Won	Currently (1965) Undecided	
I	1.05–5.86	1	3	2	6
II	0.40–1.04	1	2	1	4
III	0.18–0.39	0	1	1	2
IV	0.00–0.17	1	3	1	5
Total		3	9	5	17

282

FOOTNOTES

[1] Bruce M. Russett, *et al.*, *World Handbook of Political and Social Indicators* (New Haven, Conn.: Yale University Press, 1964).

[2] Adapted from Russett, *World Handbook*, pp. 293ff.

[3] Russett, *World Handbook*, pp. 149–57.

[4] *Ibid.*, pp. 158–61.

[5] *Ibid.*, pp. 49–55.

[6] *Ibid.*, pp. 221–26.

[7] *Ibid.*, pp. 213–16.

[8] *Ibid.*, pp. 72–81.

APPENDIX C

SUMMARIES OF WORLD WAR II UNDERGROUND RULES OF CLANDESTINE BEHAVIOR

In almost all underground movements, a set of guide rules is published for members. During World War II, these lists were used to indoctrinate underground members in the rules of clandestine and covert behavior. The following are summaries of the rules put forth by five undergrounds during World War II.

DUTCH UNDERGROUND

In the Dutch resistance, techniques of underground work were specified in detail. Members were to refrain from any activity which might draw attention to themselves or their coworkers. They were to learn as little as possible about any illegal activity beyond their own jobs. They were instructed not to tell their family or friends about anything which they were doing. Any communications which might be easily intercepted, such as mail or telephone conversations, were to be avoided. If a telephone had to be used, a public telephone should be used and a prearranged code used instead of a clear message. Members were instructed never to keep subversive literature, such as newspapers or messages, or any compromising material, on themselves or in their homes. They were to carry on a normal routine and engage in only one activity at a time.

All written material, such as lists and addresses, were to be memorized or, if this was not possible, the information was to be coded. All unnecessary contact between members was to be avoided. Agents were instructed to avoid anyone whose reliability was in doubt. The greatest single factor for the destruction of the underground organization, the directives said, was penetration by security forces. In the event of arrest, members were to remain calm and give no cause for suspicion. They were also instructed that it was imperative for them to notify their contacts in case of arrest so that the others could change their addresses and cover names and sound a warning. The Dutch underground assumed that the Germans could get whatever information they wanted through third-degree methods. However, they instructed members who might be captured to assume that the police did not know anything unless they provided unmistakable proof of guilt and to speak as little as possible so that the police could not entrap them by contradictions in their story. The worst offense that a member could make would be the betrayal of the names of others. The most important thing that he could do in the event of capture was to stall for time so that his fellow members could escape.[1]

BELGIAN UNDERGROUND

The Belgian undergrounder was instructed that during first contacts for recruiting he should feel out the sentiments of the individual to find out whether he was in favor of the occupiers or not. He was told that it was important to have an alibi wherever he went, especially when meeting someone. The members prepared a subject of conversation in advance, such that if they were captured and questioned separately that they would agree upon the alibi. If they were to meet in public, they would agree in advance on a signal to indicate danger or that it was too dangerous to talk. For meetings in homes, prearranged signals were agreed upon to warn of danger. The undergrounder was instructed to examine the meeting place before any appointment. Appointments were not to be scheduled exactly on the hour and were to be kept punctually. If meetings were scheduled between contacts who met repeatedly, one individual should leave after the other so as not to be seen in public together too frequently. Meeting places were to be varied to be sure that the individual was not followed. If he was followed, he was to walk to some isolated spot so he could check to see if he was under surveillance. He was admonished to select clothes to fit the environment and his occupation. Organizational contacts were to be limited to those within the cell or the leader or chief of the cell. All blackmail threats were to be reported immediately. He was also instructed to beware of telephone taps and postal censorship on any correspondence.[2]

DANISH UNDERGROUND

In Denmark, members were instructed to meet in parks or public places so as not to arouse suspicion. Meetings of two or three people were to be held in flats of friends or sympathizers to the movement. Members were instructed to be punctual. If captured, a member was not to give his real name but only his cover name and allow enough time for other members of his cell to flee.[3]

GERMAN ANTI-NAZI UNDERGROUND

New members were not told the exact goals and as little as possible about the organization of the movement. Seminars were conducted in which information was passed out and each new member had to apply certain views and interpretations to the information. In this manner, the individual's thinking processes were determined. Each member had a pseudonym and was instructed to change it frequently. A member received only enough information to do his job. The first five minutes of every meeting were devoted to the lesson of conspiracy: that is, members agreed upon certain facts in case the meeting was discovered and they were arrested. The story might be that a few friends had gotten together for a poker game or that they were a group of stamp collectors. One member was always directed to arrive a few minutes early or a few minutes late in order to observe the house where the meeting was taking place. If a uniformed or plain clothes policeman was in the area, the meeting would be postponed or changed to an alternate meeting place. In meeting other members, if one was more than ten minutes late, the first was to leave in order to avoid conspicuous behavior on the street or in a restaurant. Any member thought to be compromised was immediately isolated from all contacts with the organization and especially his cell. Such an individual could contact a deputy of the organization through special precautionary measures in order to reestablish contact with the organization; if he was unmolested for a long period of time, he would be taken back into the group. The conspiratorial training of the members during peacetime conditions paid off handsomely during the Nazi regime. Their organization was never infiltrated.[4]

FRENCH COMMUNIST UNDERGROUND

The Communist Party of France laid down certain rules and habits that members were expected to acquire and some habits that they were expected to rid themselves of. They were instructed that legal activities provide excellent cover for underground work. Members were warned against disclosing their identity to strangers and to beware of such shortcomings as vanity and curiosity. They were warned about complete secrecy to outsiders, and even coworkers and subordinates were only to know what they needed to know to perform their tasks. Members were to refrain from asking unnecessary and indiscreet questions of anyone within the organization. Anyone who violated this rule was to be regarded with suspicion. No meeting was to be held in which more than three members were present. No meeting was to last more than 60 minutes and the participants were expected to arrive precisely on time. Places which were likely to be under suspicion—homes of members, for example—were to be avoided in favor of such places as theater lobbies, spots in the country, or the seashore. No group was to meet at the same place twice. The plans for the meetings were never to be discussed in the mails or in the presence of third parties. The telephone was to be used only in case of emergencies. Every member who attended the meeting was to be sure that he was not followed. A member must never reveal his address, even to other members of the group. They were warned that printing and duplicating materials were not to be stored at an address known to more than two members of the group. Lists of individuals or locations were forbidden, unless in code. Members were to avoid routinized behavior, but told not to surround themselves with an atmosphere of mystery. The individual was warned that all these precautions were not easy to adopt; it would be a matter of gradually developing them over a period of time into a set of reflexes.[5]

FOOTNOTES

[1] Werner Warmbrunn, *The Dutch Under German Occupation, 1940–1945* (Stanford, Calif.: Stanford University Press, 1963), pp. 201–202.

[2] George K. Tanham, "The Belgian Underground Movement, 1940–1944" (unpublished Ph.D. dissertation, Stanford University, 1951), p. 226; also E. K. Bramstedt, *Dictatorship and Political Police: The Technique of Control by Fear* (New York: Oxford University Press, 1945), pp. 213–14.

[3] Jens Lillelund, "The Sabotage in Denmark," *Denmark During the German Occupation*, ed. Borge Outze (Copenhagen: The Scandinavian Publishing Co., 1946), pp. 50–60.

[4] Hans J. Reichhardt, "New Beginnings: A Contribution to the History of the Resistance of the Labor Movement Against National Socialism" (unpublished mimeographed manuscript, c. 1961).

[5] A. Rossi, *A Communist Party in Action* (New Haven, Conn.: Yale University Press, 1949), pp. 170–73.

GLOSSARY

Action propaganda. Immediate, observable action which follows propaganda promises: one form being specific action to alleviate hunger and suffering, thereby demonstrating insurgents' ability to accomplish set goals, and another which focuses on retaliatory acts of violence, sabotage, and punishment of so-called traitors among the local population.

AGITPROP. Communist jargon for agitation and propaganda, the principal forms of underground psychological operations. Propaganda refers to the dissemination of many ideas to a few people, usually the cadre. Agitation means the dissemination of a few ideas to the many, usually the masses.

Armed propaganda. The coordination of political propaganda with military force.

Beliefs, values, and norms. *Beliefs* are ideas, knowledge, lore, superstition, myths, and legends shared by members of a society. Associated with each cultural belief are *values*—the "right" or "wrong" judgments that guide individual actions. *Norms* are acceptable patterns of behavior which are reinforced through a system of rewards and punishments dispensed within the group.

Cadre. This term applies to the small groups of professionals or Communist vanguard who are to lead the revolution.

Cell. The basic unit of an underground organization consisting of a cell leader and cell members, with its size depending upon its specified function. The *operational cell* is usually composed of a leader and a few cell members. The *intelligence cell* is one whose leader seldom comes into contact with its members except through intermediaries such as the mail-drop, cut-out, or courier. The *auxiliary cell* is commonly found in front groups or in sympathizers' organizations, and contains an underground cell leader, assistant leaders, and members. It is structurally larger than other cells, has an intermediate level of supervision, and has little or no compartmentalization. *Parallel cells* are set up to support a primary cell and serve as backup cells.

Cells in series. Used to carry out such complex functions as the manufacture of weapons, supply, escape and evasion, propaganda, and printing of newspapers.

Chieu-hoi. The "open arms" program of South Vietnam, whereby Viet Cong defectors are offered amnesty and assistance after a short indoctrination and retraining course.

Civic Action. Any action performed by military forces of a country, using military manpower and skills, in cooperation with civil agencies, authorities, or groups, that is designed to improve the economic or social betterment of that country. Civic action programs are designed to enhance the stature of indigenous military forces and improve their relationship with the population.

Clandestine operations. Activities to accomplish intelligence, counter-intelligence, and other similar activities in such a way as to assure secrecy or concealment.

Cominform. The Communist Information Bureau.

Comintern. The Communist International.

Consensual validation. The device of confirming facts or norms through group approval or consensus.

CONTACT INTELLIGENCE. Immediately usable intelligence which provides information as to the immediate whereabouts and identity of a subversive individual or group. There are essentially four methods of obtaining contact intelligence: 1. *Patrols.* Patrols or police squads search for physical evidence such as tracks or campsites, and for consistent patterns in enemy movements. 2. *Low-level informants.* These members look for even the smallest tips or leads and bits of information that might help to complete the background information on the insurgents. 3. *Forced contact.* An example of this tactic might be cutting off the guerrillas from their underground supply source, thereby forcing them into the open to contact their support arm. 4. *Informants.* The use of informants to gain intelligence about the movement, position, and activity of insurgents.

Cooptation. The practice of utilizing the special talents or qualifications of individuals who may be indifferent or opposed to the goals of the organization by giving them a position of nominal importance on the periphery of the organization.

Cordon-and-search technique. A method of gathering intelligence when the populace does not cooperate for fear of reprisal from the underground. Usually the security forces seal off the entrances and escape routes and search all people and property within the area.

Counterinsurgency. Those military, paramilitary, political-economic, psychological, and civic actions taken by a government to defeat subversive insurgency.

Covert operation. Operations which are so planned and executed as to conceal the identity of or permit plausible denial of subversive operations. They differ from clandestine operations in that emphasis in clandestine operations is placed on concealment of the operation rather than the concealment of personal identity.

Criticism, self-criticism. The actual activities of criticism and self-criticism sessions consist of conferences, discussions, and meetings within the party in which personal performance is evaluated by the individual and the group to determine and correct any weaknesses in the work of the party or party members.

Double language. A technique by which instructions to the Communist cadre are concealed in propaganda materials.

EDCOR. In the Philippines, the government's Economic Development Corps.

Escape-and-evasion nets. A system established by undergrounds for purposes of infiltration and exfiltration of the area of hostile operations.

Fail-safe principle. Principle by which if one element or operation fails or is compromised the consequences to the overall organization or operation will be minimal.

Front organization. Commonly refers to political activities carried out behind the facade of an apparently non-Communist organization.

GRU. The Soviet Military Intelligence Directorate.

Guerrilla force. The relatively small visible element of a revolutionary movement organized to perform overt armed military and paramilitary operations using guerrilla tactics.

Invulnerability concept. The practice of frequent assignments and a high degree of activity have the useful side effects of keeping the individual so engrossed in his work that he loses any fear of harm coming to him, and unconsciously considers himself invulnerable.

KGB. The Soviet Committee for State Security.

Mail-drop. A mail-drop is placed where a message may be left by one person to be picked up later by another.

Main force, regional force, and local force. The military elements of an insurgency. The regular main force is organized along conventional military lines, such as platoons, etc.; the regional troops have responsibility for an area comparable to a province or state; and the local militia is composed of villagers operating at the village level. Generally the main force uses conventional tactics while the regional and local militia use guerrilla tactics.

Passive resistance. This method implies a large unarmed group whose activities capitalize upon social norms, customs, and taboos in order to provoke action by security forces that will serve to alienate large segments of public opinion from the government or its agents.

Politbureau. The principal policymaking and executive committee of the Communist Party.

Population and resources control. That aspect of the counterinsurgency effort designed to control human and material resources. Objectives of this effort are to sever the relationship between the population and the guerrilla; identify and neutralize the insurgent apparatus and activities within the population; and create, within the population, a secure physical and psychological environment.

Polygraph. A device which measures cardiovascular, respiratory, and galvanic skin responses, used extensively in criminal investigations and interrogation in counterinsurgency operation.

Pseudogangs. A team of infiltrators highly trained and indoctrinated in local mannerisms, attitudes, speech and dress, down to the minutest detail, to simulate an insurgent group for purposes of infiltration and intelligence collection.

Revolutionary or insurgent movement. A subversive, illegal attempt by an organized indigenous group outside the established governing structure to weaken, modify, or replace an existing government through the protracted use or threatened use of force.

Sabotage. An attempt by insurgents to withhold resources from the government's counterinsurgency effort by acts of destruction.

Safe houses. Hideouts which are part of an escape-and-evasion network.

Selective exposure. The tendency on the part of individuals to hear only information congenial to their own tastes, biases, and existing attitudes.

Selective interpretation. Information understood only in terms of prior attitudes.

Surveillance. The covert observation of persons and places, including mechanical observation such as wiretaps or concealed microphones.

Terrorism, or terror. Terrorism is those coercive acts of violence utilized by a subversive movement and usually directed toward disrupting government control over the citizenry and creating a state of mind—terror—which makes the citizenry acquiesce to subversive demands.

Underground. The clandestine or covert organizational element of an insurgent movement.

United front. A common Communist tactic which creates an alliance against the government of all organizations or forces of discontent.

☆ U.S. GOVERNMENT PRINTING OFFICE: 1966-O 225-345